BASIC
PLUMBING

BASIC PLUMBING

THOMAS PHILBIN

RESTON PUBLISHING COMPANY, INC.
Reston, Virginia
A Prentice-Hall Company

Library of Congress Cataloging in Publication Data

Philbin, Thomas
  Basic plumbing.

  Includes index.
  1. Plumbing. I. Title.
TH6124.P47   696'.1   76-41166
ISBN 0-87909-065-0

© 1977 by
Thomas Philbin

10 9 8 7 6 5 4 3 2

Printed in the United States of America.

Contents

I wrote this book as a basic text for the student of plumbing, but there's no reason why the home handyman can't put it to good use. Indeed, I think there is one good reason why it should be used by homeowners. There are many jobs described here that are well within the capability of the average or even the beginning handyman. Most repairs, for example, are in this category. House plumbing problems are usually relatively minor: a faucet leaks, a toilet doesn't stop running, a sink clogs. Once you know how, it's simple to fix these problems.

But the handyman can also do plumbing home improvements—adding a sink, a toilet, a bath. Since the book is designed to give the student plumber detailed information on such jobs, it will let the handyman go as far as he is able to go.

Even if you don't do every job in the book (and, of course, you won't), you will at least get an overview of the complete plumbing system of your home. It might not be crucial to have a complete picture of this system when you're fixing a leaky faucet, but there's something reassuring about being familiar with it.

I have always been a great believer that illustrations are better than words for conveying how-to information, and I think this book shows that. There are hundreds of drawings and photographs showing a wide variety of jobs in step-by-step fashion. And to help you figure the amount of materials you need for a job or to make other important mathematical computations related to plumbing, a chapter on "Mathematics Down the Drain" by Nancy Walordy, a former math and science teacher, will help you review basic arithmetic and geometry.

At any rate, good luck to the student plumber and the handyman.

Thomas Philbin

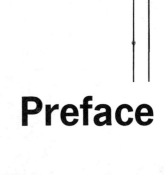

Preface

BASIC PLUMBING

1

Just as a doctor must understand the whole of the human body before he can treat a part of it properly, it is important for the plumber to understand how the plumbing system of a home works. In general, the plumbing system is composed of three systems, or parts: the fresh water supply, the drainage system, and the fixtures and appliances.

FRESH WATER SUPPLY

Fresh water normally has one of two sources. Either it starts its journey to the house from a public water works, or from a well.

Water that comes from a water works is first purified there and is sent under pressure through pipes to the water service line, which carries the water directly into the house.

When it enters the house, the water travels through a meter, which records how much is used; then it travels by means of a single pipe to the hot water heater. Here, a short branch line carries some of the water into the heater. When the water is heated, and as demand for it is made by the fixtures and appliances, it travels out of another pipe from the water heater, then travels parallel with the cold water line throughout the house to the various fixtures and appliances. The hot water line is always on the left (as you face it) and the cold water line on the right.

Various names have been given to the water supply system pipes. Those that rise vertically at least one floor through the house are called *risers;* those

How a House Plumbing System Works

1

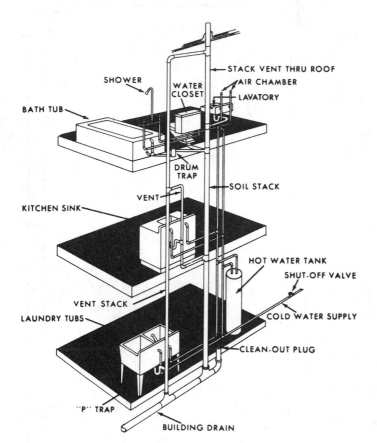

FIGURE 1-1

A basic plumbing system, showing the water and drain-waste-vent (DWV) systems.

that branch off with water directly to fixtures are called *branch lines.*

At various points in the system there are *valves,* which are devices for turning off the water, either hot or cold, when repairs are necessary. Valves are located on the pipes so that they control certain portions. If repair is required you can turn off only the affected portions, rather than the whole system. Valves are also placed at the bases of all risers. This facilitates emptying the system of water for winterizing.

In municipally supplied water systems there is a valve that well-supplied systems do not have. This is located on the line that comes off the main line that passes a home. It is opened and closed with a long-handled wrench called a "street key." When the cold water supply is shut off, the hot water supply throughout is also off.

There are one or two main valves located at the water meter. You can turn these by hand, thereby shutting down all the water in the house. One is located on the far side of the meter, and is called the *meter shutoff valve,* and the other on the near side (or house side) is the *main shutoff*

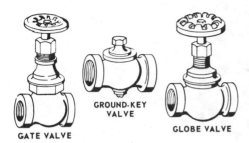

GATE VALVE   GROUND-KEY VALVE   GLOBE VALVE

FIGURE 1-2

Three kinds of valves commonly found in residential plumbing.

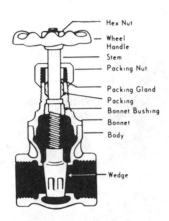

FIGURE 1-3

Cross sectional view of a gate valve, the most efficient type. When the wedge is lifted by turning the handle, there is almost unrestricted water flow.

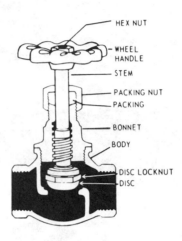

FIGURE 1-4

Cross section view of a globe valve. This type is less efficient than the gate valve.

(Fig. 1-5). When shutting down house water, always turn off the main, rather than the meter, valve.

There are also usually valves (and they are desirable) under individual fixtures. Hot water heaters and water treatment equipment also usually have shutoff valves, and it is a good idea to know where these are.

Another important part of the water supply system is the *air chambers*—although not all systems have them. These are vertical lengths of pipe with metal caps at the top. The chambers extend from water supply pipes at the point where the pipes enter the fixture. Air chambers are shock absorbers. When water pressure builds up to amounts that are too much for the particular size of pipe to handle, these air chambers fill with the "excess" water so that the pipes do not start rocking and rolling, banging

FIGURE 1-5
The main shutoff valve just inside the house. Turning this valve shuts down water in the entire house.

FIGURE 1-6
Another shutoff, this one is located outside the house. The job is done with pliers.

loudly (a malady called "water hammer"), or perhaps shaking to the point of breakage.

FIXTURES

The sink, the tub, and the toilet—the fixtures—are the main units where the water is used. In a sense, even the faucets on the outside of a house can be considered fixtures, although they are more properly called *fittings*. We need not discuss these at length now, but you should know that they are considered part of a plumbing system.

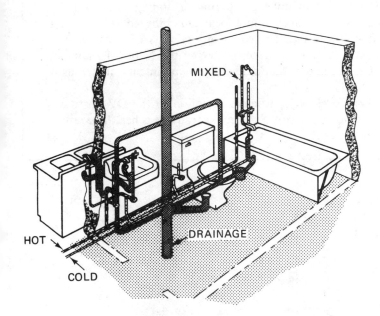

FIGURE 1-7

For economy of installation, it is a good idea to locate the plumbing system so that fixtures can be installed back to back on both sides of a wall.

THE DRAINAGE SYSTEM

The third component in a plumbing system is the drainage system, more properly called the DWV or drain-waste-vent system, to describe the functions it performs: draining away wastes, and venting gases at the same time. Generally, the system consists of the soil stack, the waste pipes, the traps, and the vents.

In operation, waste first runs out of the fixture, say a sink. From there it enters the waste pipes, usually 1½-inch to 2-inch diameter pipes, which are sloped and lead to the soil stack. The stack is a large-diameter pipe that runs vertically from the lowest point in the plumbing system to above the roof and out the roof. In older homes, 4-inch cast iron pipe is most commonly used. In modern homes, you will find 3-inch copper tube and 4-inch ABS (acrylonitrile-Butadiene styrene) and PVC (polyvinyl chloride) plastic pipe used. The waste then runs down the soil stack, or *stack,* as it is commonly called, which in turn leads to the building drain, then to the sewer line, which leads to the city sewer line, cesspool, or septic system.

THE PLUMBING CODE

For work that involves altering or adding to a plumbing system, the plumber must follow local plumbing codes as to materials and workmanship. These can vary greatly. Some are old and restrictive; others more modern, following the National Plumbing Code; still others allow what amounts to experimental materials to be used.

Most codes pay particular attention to the DWV system, because if installed improperly it could result in a health hazard—sewage could get into the water supply system.

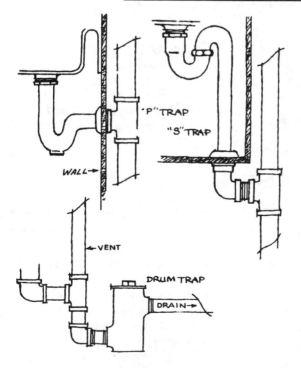

FIGURE 1-8
Three kinds of traps.

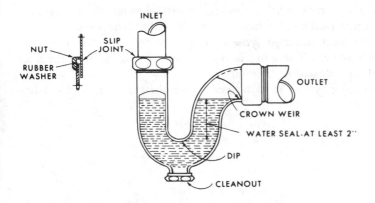

FIGURE 1-9

Cross sectional view of a
P-trap. The water seal on any
trap should be at least 2".

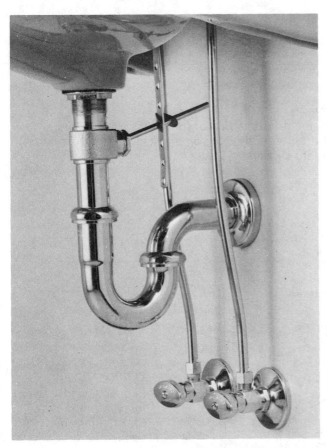

FIGURE 1-10

A P-trap and water supply lines
with valves at the bottom.

As a safety measure against gases backing up into the house, all
fixtures are equipped with traps (vari-shaped pipe sections) directly under
them, where water is trapped (hence the name *trap*). These provide a seal

against gases and vermin getting into the house through drains. Toilets have built-in traps (toilets also have waste pipes that lead directly to the soil stack). The most common types of traps are "P" and "S" types, so named for their shapes. There are also a house trap and a fresh air inlet where the soil pipe leaves the building.

A plumbing system must be vented for three reasons: (1) it lets

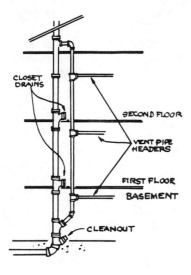

FIGURE 1-11

A basement-to-roof view of the DWV system. Toilets must drain directly into soil stacks; other fixtures can drain into secondary stacks.

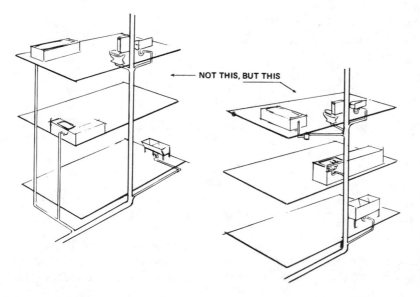

FIGURE 1-12

Economy and ease of installation dictate that fixtures be connected to a single stack, which drains wastes and vents gases, among other things.

potentially dangerous sewer gases escape; (2) it equalizes air pressure in the system so that waste and water can flow freely; (3) it prevents the

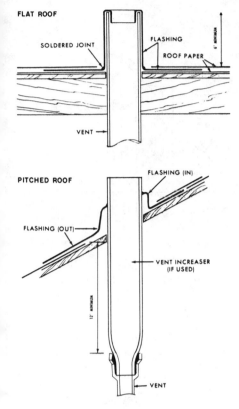

FIGURE 1-13
Two methods of flashing the stack when it exits through the roof.

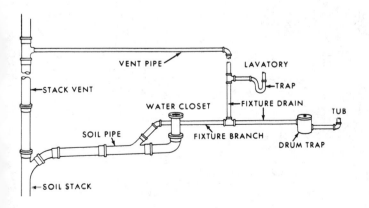

FIGURE 1-14
A closeup view of part of the DWV system. Soil pipes only handle human waste directly. The bottom part of the soil stack handles liquid waste, while the top part is designated as a vent.

backup of water into a fixture that is located below another fixture that happens to be draining.

Every fixture in the house is connected either to the main stack, or to smaller stacks (where the pipe would have to be run too far to reach

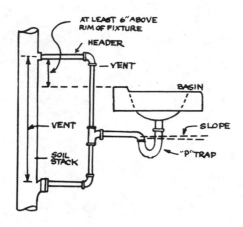

FIGURE 1-15
A typical installation of a lavatory to the soil stack.

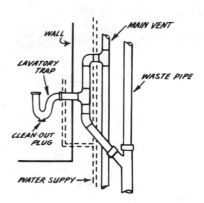

FIGURE 1-16
Another method of connecting a lavatory to the soil stack.

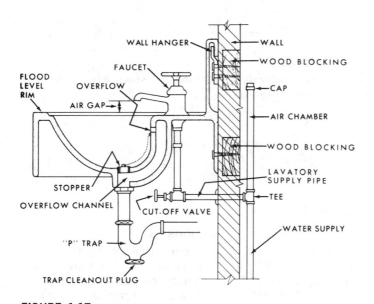

FIGURE 1-17
A cross sectional view of a lavatory, showing how it is tied into water lines.

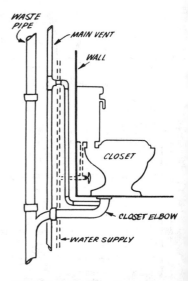

FIGURE 1-18
Tying a water closet into the system.

the main stack) that project through the roof: the entire system is open to the outside air. The stack itself serves a double purpose: the bottom section below the fixtures is a waste line, and the portion above is designated as a vent.

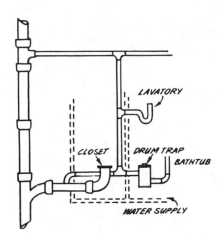

FIGURE 1-19
An example of all three bathroom fixtures tied into the system.

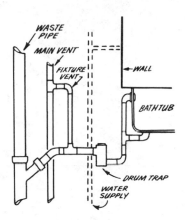

FIGURE 1-20
Connecting the main vent to a fixture vent.

At the base of each soil stack there is also a cleanout plug. And handy they are. If, for example, you cannot clean out a drain line blockage by sticking a snake down a sink drain, you can attack it lower down in the pipe by just removing the cleanout plug and inserting your snake in there. Cleanouts should also be located at each change in direction of the waste line, where the chance is greater that blockage will occur.

2

In addition to what can be regarded as basic plumbing tools—wrenches, pipe cutters, etc.—the plumber needs a variety of other tools to do his job properly. Following is a roundup of both kinds of tools, combined with tips on their use. In succeeding chapters, additional details on their use are shown in relation to specific jobs, such as "Installing a Drainage System," Chapter 6.

CARPENTER'S HAMMER

The primary use of the carpenter's hammer is to drive or draw (pull) nails. The hammer has either a curved or a straight claw. The face may be either bell-faced, or plain-faced, and the handle may be made of wood or steel.

The ball-peen hammer has a ball that is smaller in diameter than the face and is useful for striking areas that are too small for the face to enter. Ball-peen hammers are made in different weights, usually 4, 6, 8, and 12 ounces and 1, 1½, and 2 pounds. For most work a 1½-pound hammer or a 12-ounce hammer is good. However, a 4- or 6-ounce hammer is often used for light work, such as tapping a punch to cut gaskets out of sheet gasket material.

using hammers

Simple as the hammer is, there is a right way and a wrong way of using it. The most common fault is holding the handle too close to the head, or *choking;* it reduces the force of the blow. It also makes it harder to hold the head in an upright position. Except for light blows,

Tools
of the Trade

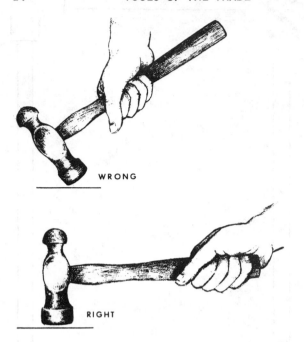

WRONG

RIGHT

FIGURE 2-1

The right and wrong ways to strike a surface with a ball-peen hammer.

hold the handle close to the end with four fingers underneath and the thumb along the side or on top of the handle. The thumb should rest on the handle and never overlap the fingers.

Try to hit the object with full force, holding the hammer at such an angle that the face and the surface being hit are parallel. This distributes the force of the blow over the full face and prevents damage to the surface and the hammer face.

To drive a nail, hold it down near the point and set it on the surface, holding it straight. Give it a few light taps to get it started, then follow with full-force blows. If you do not want to risk damaging the surface of the work with a blow, you can drive the nail to within a fraction of the surface, then drive it the rest of the way with a nail set.

You can pull small nails by simply slipping the claw of the hammer over the head and prying backwards. For large nails, however, you can get extra leverage if you place a small block of wood under the hammer before pulling back.

maintenance

Hammers should be cleaned and repaired, if necessary, before they are stored. Before using, insure that the faces are free from oil or other material that would cause the tool to glance off nails, spikes, or stakes. The heads should be dressed to remove any battered edges.

The hammer handle should always be tight in the head. If it is loose, the head may fly off and cause an injury. The eye or hole in the hammer head is made with a slight taper in both directions from the center. Once the handle, which is tapered to fit the eye, is inserted in the head, a steel or wooden wedge is driven into the end of the handle that is inserted in the head. This wedge expands the handle and causes it to fill the opposite taper in the eye. Thus the handle is wedged in both directions. If the wedge starts to come out, drive it in again. If the wedge comes out, replace it before continuing to use the hammer. If you cannot get another wedge right away, you may file one out of a piece of flat steel, or cut one from a portion of the tang of a wornout file. (The tang is the end of the file that fits into the handle.)

safety

Hammers are dangerous tools when used improperly. Some important things to remember:

1. Do not use a hammer handle for bumping parts in assembly, and never use it as a pry bar. Such abuses cause the handle to split; a split handle can produce bad cuts or pinches. When a handle splits or cracks, do not try to repair it by binding with string or wire; replace it.
2. Make sure that the handle fits tightly on the head.
3. Do not strike a hardened steel surface. Small pieces of steel may break off and injure someone in the eye or damage the work. It is all right to strike a punch or chisel directly with a ball-peen hammer, because the steel in the heads of punches and chisels is slightly softer than that in the hammerhead.

WRENCHES

A wrench is a basic plumbing tool that is used to exert a twisting force on bolts, nuts, studs, and pipes.

The best wrenches are made of chrome vanadium steel. Wrenches made of this material are lightweight and almost unbreakable. Most common wrenches are made of forged carbon steel or molybdenum steel. These latter materials make good wrenches, but they are generally made a little heavier and bulkier in order to achieve the same degree of strength as chrome vanadium tools.

The size of any wrench used on bolt heads or nuts is determined by the size of the opening between the jaws of the wrench. The opening is manufactured slightly larger than the bolt head or nut that it is designed to fit. Hex-nuts (six-sided nuts) and other types of nuts or bolt heads are measured across opposite flats (surfaces). A wrench that is designed to fit a ⅜-inch nut or bolt usually has a clearance of from 5 to 8 thousandths of an inch. This clearance allows the wrench to slide on and off the nut or bolt with a minimum of "play." If the wrench is too large, the points of the nut or bolt head can be rounded and destroyed.

There are many types of wrenches. Each type is designed for a specific use. Let us discuss some of them.

open-end wrenches

Solid, nonadjustable wrenches with openings in one or both ends are called open-end wrenches. Usually they come in sets of from six to ten wrenches with sizes ranging from ⁵⁄₁₆ to 1 inch. Wrenches with small openings are usually shorter than wrenches with large ones. This proportions the lever advantage of the wrench to the bolt, stud, or pipe and helps prevent wrench breakage or damage to the item.

Open-end wrenches may have their jaws parallel to the handle or at angles of up to 90 degrees. The average angle is 15 degrees. This variation permits selection of a wrench suited for places where there is room to make only a part of a complete turn of a nut or bolt. If the wrench is turned over after the first swing, it will fit on the same flats and turn the nut farther. After two swings on the wrench, the nut is turned far enough so that a new set of flats is in position for the wrench.

Handles are usually straight, but may be curved. Those with curved handles are called S-wrenches. Other open-end wrenches may have offset

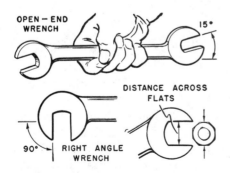

FIGURE 2-2
Fixed open-end wrenches come in various shapes.

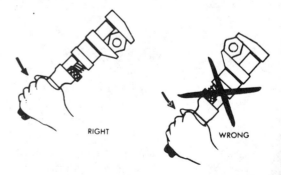

FIGURE 2-3
The right and wrong ways to apply turning force to a wrench.

handles, which allow the head to reach nut or bolt heads that are sunk below the surface.

box wrenches

Box wrenches are safer than open-end wrenches, since there is less likelihood that they will slip off the work. They completely surround or "box" a nut or bolt head.

The most frequently used box wrench has 12 points or notches arranged in a circle in the head and can be used with a minimum swing angle of 30 degrees. Six- and eight-point wrenches are used for heavy-duty work, 12-point wrenches for medium work, and 16-point wrenches for light-duty work only.

One advantage of 12-point construction is the thin wall. It is better for turning nuts that are hard to get at with an open-end wrench. Another advantage is that the wrench will operate between obstructions where the space for handle swing is limited. A very short swing of the handle will turn the nut far enough to allow the wrench to be lifted and the next set of points to be fitted to the corners of the nut.

One disadvantage of the box-end wrench is the loss of time that occurs whenever the plumber has to lift the wrench off and place it back on the nut.

combination wrenches

After a tight nut is broken loose, it can be unscrewed much more quickly with an open-end wrench than with a box wrench. This is where a combination box-open-end wrench comes in handy. You can use the box end for breaking nuts loose or for snugging them down, and the open end for faster turning.

The box-end portion of the wrench can be designed with an offset in the handle.

Correct use of open-end and box-end wrenches can be summed up in a few simple rules, most important of which is to be sure that the wrench properly fits the nut or bolt head.

When you have to pull hard on the wrench, as in loosening a tight nut, make sure that the wrench is seated squarely on the flats.

Pull on the wrench—do not push. Pushing a wrench is a good way to skin your knuckles if the wrench slips or the nut breaks loose unexpectedly. If it is impossible to pull the wrench, and you must push, do it with the palm of your hand and hold your palm open.

Only practice will tell you if you are using the right amount of force on the wrench so that you can turn the nut to proper tightness without stripping the threads or twisting off the bolt.

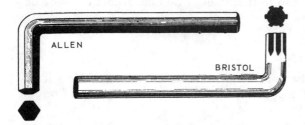

FIGURE 2-4

Allen and bristol wrenches open specific-shaped nuts. The Allen wrench, which comes in various sizes, is particularly useful for removing faucet seats (see Chapter 10).

adjustable wrenches

A handy all-around wrench that is a must for your toolbox is the adjustable wrench. These have jaws that can be opened or closed. Within the limitations of the individual wrench, they fit nuts, bolts, and pipes of various sizes. There are many different types of adjustable wrenches, several of which are used in plumbing.

adjustable open-end wrench

Use this type of wrench on square or hexagonal nuts when you are working with the interior parts of faucets and valves. For most minor plumbing jobs, a 12-inch wrench is probably the best choice. (The size refers to the length of the tool; however, the jaws of larger wrenches do open more widely.) It is also handy to have an 8-inch or 6-inch wrench for smaller work.

monkey wrench

This type of wrench is used for the same purposes and in the same way as the adjustable open-end wrench. Turning force should be applied to the back of the handle (the side of the wrench opposite the jaw opening).

pipe wrenches

For rotating or holding round work an adjustable pipe wrench (Stilson) may be used. The movable jaw on a pipe wrench is pivoted to permit a gripping action on the work. This tool must be used with discretion, as the jaws are serrated and always make marks on the work unless adequate precautions are observed. The jaws should be adjusted so that the bite on the work will be taken at about the center of the jaws.

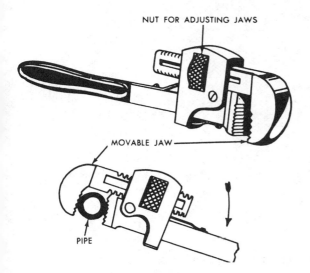

NUT FOR ADJUSTING JAWS

MOVABLE JAW

PIPE

FIGURE 2-5
Pipe and Stillson wrenches are used to turn pipe.

chain pipe wrenches

A different type of pipe wrench, used mostly on large sizes of pipe, is the chain pipe wrench. This tool works in one direction only, but it can be backed partly around the work and a fresh hold can be taken without freeing the chain. To reverse the operation the grip is taken on the opposite side of the head. The head is double-ended and can be reversed when the teeth on one end are worn out.

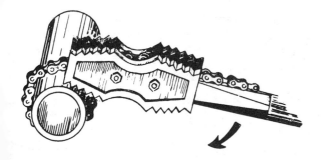

FIGURE 2-6
A chain pipe wrench, good for handling large-diameter pipe.

strap wrenches

The strap wrench is similar to the chain pipe wrench, but it uses a heavy web strap in place of the chain. This wrench is used for turning pipe when you do not want to damage the pipe.

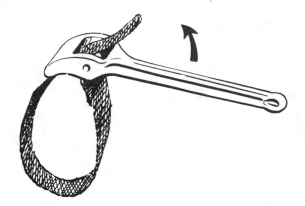

FIGURE 2-7
A strap wrench is handy when you don't want to damage pipe.

rules for safety

1. Always use a wrench that fits the nut properly.
2. Keep wrenches clean and free from oil. Otherwise they may slip, and serious injury to you or damage to the work may result.
3. Do not increase leverage of a wrench by placing a pipe over the handle. Increased leverage may damage the wrench or the work.
4. Provide some sort of kit or case for all wrenches. Return them to it at the completion of each job. Doing so saves time and trouble and facilitates selection of tools for the next job. Most important, it eliminates the possibility of leaving them where they can cause injury or damage to men or equipment.
5. Determine the way in which a nut should be turned before trying to loosen it. Most nuts are turned counterclockwise for removal. This may seem obvious, but even experienced men have been observed straining at the wrench in the tightening direction when they wanted to loosen it.

METAL-CUTTING TOOLS

There are many types of metal-cutting tools, a few of which (snips and shears) are good for plumbing jobs. These are used for cutting sheet metal and steel of various thicknesses and shapes. Normally, the heavier or thicker material is cut by shears.

snips

One of the handiest tools for cutting light (up to $\frac{1}{16}$ inch thick) sheet metal is the hand snip or tin snips. The straight hand snips shown in Fig.

2-8 have blades that are straight and cutting edges that are sharpened to an 85-degree angle. Snips like these can be obtained in different sizes ranging from small 6-inch to large 14-inch snips. Tin snips also work on slightly heavier gauges of soft metals, such as aluminum alloys.

STRAIGHT HAND SNIPS

CIRCLE SNIPS

HAWKS-BILL SNIPS

TROJAN SNIPS

AVIATION SNIPS

FIGURE 2-8

Snips.

    Snips do not remove any metal when a cut is made. There is danger, however, of causing minute metal fractures along the edges of the metal during the cutting. For this reason, it is better to cut just outside the layout line. This procedure allows you to dress the cutting edge while keeping the material within required dimensions.

    Cutting extremely heavy-gauge metal always presents the possibility of springing the blades. Once the blades are sprung, hand snips are useless. If you use the rear portion of the blades for such cutting, you not only avoid the possibility of springing the blades but also obtain greater cutting leverage.

    Many snips have small serrations on the blades. These serrations tend to prevent the snips from slipping backwards when a cut is being made. Although this feature does make the cutting easier, it mars the edges of the metal slightly. You can remove these small cutting marks if you allow proper clearance for dressing the metal to size. There are many other types of hand snips used for special jobs, but the snips discussed can be used for almost any common type of work.

cutting sheet metal with snips

    It is hard to cut circles or small arcs with straight snips. There are snips especially designed for circular cutting. They are called circle snips, hawks-bill snips, trojan snips, and aviation snips.

    To cut large holes in the lighter gauges of sheet metal, start the cut by punching or otherwise making a hole in the center of the area to be cut out. With aviation snips or some other narrow-bladed snips, make a spiral cut from the starting hole out toward the scribed circle and continue cutting until the scrap falls away.

To cut a disk in the lighter gauges of sheet metal, use combination snips or straight-blade snips. First, cut away any surplus material outside of the scribed circle, leaving only a narrow piece to be removed by the final cut. Make the final cut just outside the layout line. This will let you see the line while you are cutting and will cause the scrap to curl up below the blade during cutting.

FIGURE 2-9
When cutting material from the inside of sheet metal, start with a hole in the middle.

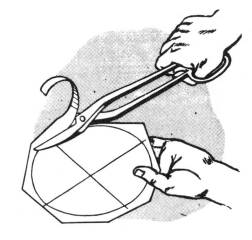

FIGURE 2-10
Trimming a sheet metal circle with snips.

To make straight cuts, place the metal on a bench with a marked guide line over the edge of a workbench. Hold the sheet down with one hand. With the other hand, hold the snips so that the flat sides of the blades are at right angles to the surface of the work, and cut. If the blades are not at right angles, the edges of the cut will be slightly bent and burred. The bench edge should act as a guide when you are cutting.

Any of the hand snips may be used for straight cuts. When notches are too narrow to be cut out with a pair of snips, make the side cuts with the snips and cut the base of the notch with a cold chisel.

safety and care

Snips should always be oiled and adjusted to permit easy cutting and a cut that is free from burrs. If blades bind, or if they are too far apart, the snips should be adjusted by tightening the screw that holds them together.

Never use snips as screwdrivers, hammers, or pry bars. They break easily.

Do not attempt to cut heavier materials than the snips are designed for. Never use tin snips, for example, to cut hardened steel wire or other similar objects. Such use will dent or nick the cutting edges.

Never toss snips into a toolbox where the cutting edges can come into contact with other tools. Such contact dulls the cutting edges and may even break the blades. When snips are not in use, hang them on hooks or lay them on an uncrowded shelf or bench.

hacksaws

Hacksaws are used to cut metal that is too heavy for snips; for example, metal bar stock can be cut readily with hacksaws. They are also fundamentally important in cutting many types of pipe.

There are two parts to a hacksaw: the frame and the blade. Common hacksaws have either an adjustable or a solid frame. Adjustable frames can be made to hold blades from 8 to 16 inches long, whereas those with solid frames take only the blade of the length for which they are designed. This length is the distance between the two pins that hold the blade in place.

Hacksaw blades are made of high-grade tool steel, hardened and tempered. There are two types, the all-hard and the flexible. All-hard blades are hardened throughout, whereas only the teeth of the flexible blades are hardened. Hacksaw blades are about one-half inch wide, have from 14 to 32 teeth per inch, and are from 8 to 16 inches long. The blades have a hole at each end that hooks to a pin in the frame. All hacksaw frames that hold the blades either parallel or at right angles to the frame are provided with a wing nut or screw to permit tightening or removing the blade.

The *set* in a hacksaw refers to how much the teeth are pushed out in opposite directions from the sides of the blade. The four different kinds of set are alternate set, double alternate set, raker set, and wave set.

The teeth in the alternate set are staggered one to the left and one

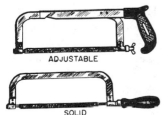

ADJUSTABLE

SOLID

FIGURE 2-11
Two types of hacksaws.

to the right throughout the length of the blade. On the double alternate set blade, two adjoining teeth are staggered to the right, two to the left, and so on. On the raker set blade, every third tooth remains straight and the other two are set alternately. On the wave (undulated) set blade, short sections of teeth are bent in opposite directions.

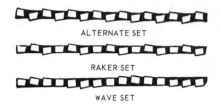

ALTERNATE SET

RAKER SET

WAVE SET

FIGURE 2-12

Types of hacksaw teeth.

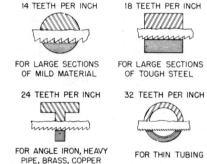

14 TEETH PER INCH

FOR LARGE SECTIONS OF MILD MATERIAL

18 TEETH PER INCH

FOR LARGE SECTIONS OF TOUGH STEEL

24 TEETH PER INCH

FOR ANGLE IRON, HEAVY PIPE, BRASS, COPPER

32 TEETH PER INCH

FOR THIN TUBING

FIGURE 2-13

Hacksaw blades must be matched to the material being cut.

using hacksaws

The hacksaw is often used incorrectly. Although it can be used with limited success by an inexperienced person, a little thought and study given to its proper use will result in faster and better work and less dulling and breaking of blades.

Good work with a hacksaw depends not only on proper use of the saw but on proper blade selection. Coarse blades with fewer teeth per inch cut faster and are less liable to choke up with chips. However, finer blades with more teeth per inch are necessary when thin sections are to be cut.

To make the cut, first install the blade in the hacksaw frame so that the teeth point away from the handle. (Hand hacksaws cut on the push stroke.) Tighten the wing nut so that the blade is under tension; this helps make straight cuts.

Place the material to be cut in a vise. A minimum of overhang reduces vibration, gives a better cut, and lengthens the life of the blade. Have the layout line outside of the vise jaw so that the line is visible while you work.

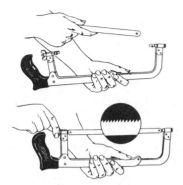

FIGURE 2-14

Inserting the blade in an adjustable hacksaw. Note the direction of the teeth.

The proper method of holding the hacksaw is shown in Fig. 2-15. Note how the index finger of the right hand, pointed forward, aids in guiding the frame.

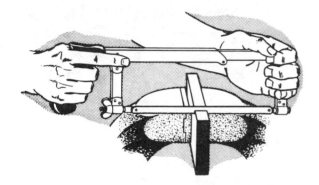

FIGURE 2-15

The proper way to hold a hacksaw when cutting.

When cutting, let your body sway ahead and back with each stroke. Apply pressure on the forward stroke, which is the cutting stroke, but not on the return stroke. From 40 to 50 strokes per minute is the usual speed. Long, slow, steady strokes are preferred.

For long cuts rotate the blade in the frame so that the length of the cut is not limited by the depth of the frame. Hold the work with the layout line close to the vise jaws, raising the work in the vise as the sawing proceeds.

Saw thin metal as shown in Fig. 2-16. Notice the long angle at which the blade enters the saw groove or kerf. This permits several teeth to be cutting at the same time.

Metal that is too thin to be held can be placed between blocks of wood. The wood provides support for several teeth as they are cutting. Without the wood, teeth will be broken because of excessive vibration of the stock and because individual teeth must absorb the full power of the stroke. Pipe may also be placed in a miter box for cutting.

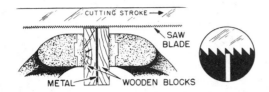

FIGURE 2-16

The proper method for cutting thin metal with a hacksaw.

Cut thin metal with layout lines on the face and a piece of wood behind. Hold the wood and the metal in the jaws of the vise, using a C-clamp when necessary. The wood block helps support the blade and produces a smoother cut. Using the wood only in back of the metal permits the layout lines to be seen.

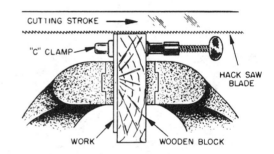

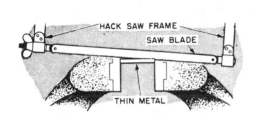

FIGURE 2-17

Two ways to cut thin metal with a hacksaw.

To remove a frozen nut with a hacksaw, saw into the nut, starting the blade close to the threads on the bolt or stud and parallel to one face of the nut. Saw parallel to the bolt until the teeth of the blade almost reach the lockwasher. Lockwashers are hard and will ruin hacksaw blades, so do not try to saw them. Then, with a cold chisel and hammer, remove this one side of the nut completely by opening the saw kerf. Put an adjustable wrench across this new flat and the one opposite, and again try to remove the frozen nut. Since very little original metal remains on this one side of

FIGURE 2-18

Cutting a piece off the edge of the work with a hacksaw.

the nut, the nut will either give or break away entirely and permit its removal.

To saw a wide kerf in the head of a cap screw or machine bolt, fit the hand hacksaw frame with two blades side by side, and with teeth lined up in the same direction. With slow, steady strokes, saw the slot approximately one-third the thickness of the head of the cap screw. Such a slot will permit subsequent holding or turning with a screwdriver when it is impossible, due to close quarters, to use a wrench.

hacksaw safety

The main danger in using hacksaws is injury to your hand if the blade breaks. The blade will break if too much pressure is applied, when the saw is twisted, when the cutting speed is too fast, or when the blade becomes loose in the frame. Additionally, if the work is not tight in the vise, it will sometimes slip, twisting the blade enough to break it.

rod saws

An improvement in industrial technology provides us with a tool that can cut material that an ordinary hacksaw cannot even scratch. The rod saw acts like a diamond, cutting hard metals and materials such as stainless steel, glass, and stone. It cuts by means of hundreds of tungsten-carbide particles permanently bonded to the rod.

A unique feature of this saw is its capability of cutting on the forward and reverse strokes.

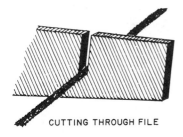

CUTTING THROUGH FILE

FIGURE 2-19
Closeup view of a rod saw blade. It can cut everything from glass to hard metal.

chisels

Chisels are tools that can be used for chipping or cutting metal (or wood). They will cut any metal that is softer than the material that they are made of. Chisels are made from a good grade of tool steel and have a hardened cutting edge and beveled head. Cold chisels are classified according to the shape of their points; the width of the cutting edge indicates their size. The most common shapes of chisels are flat (cold chisel), cape, round nose, and diamond point.

The kind of chisel most commonly used is the flat cold chisel, which serves to cut rivets, to split nuts, to chip castings, and to cut thin metal sheets and cast-iron pipe. The cape chisel is used for special jobs like cutting keyways, narrow grooves, and square corners. Round-nose chisels make circular grooves and chip inside corners with a fillet. Finally, the diamond-point chisel is used for cutting V-grooves and sharp corners.

FIGURE 2-20
The cold chisel is an invaluable tool for cutting cast iron soil pipe.

using chisels

As with other tools, there is a correct technique for using a chisel. Select a chisel that is large enough for the job. Be sure to use a hammer that matches the chisel; that is, the larger the chisel, the heavier the hammer should be. A heavy chisel will absorb the blows of a light hammer and will do virtually no cutting.

As a general rule, hold the chisel in the left hand with the thumb and first finger about one inch from the top. It should be held steadily but not tightly. The finger muscles should be relaxed, so that if the hammer strikes the hand, the hand can slide down the tool, lessening the effect of the blow. Keep the eyes on the cutting edge of the chisel, not on the head, and swing the hammer in the same plane as the body of the chisel. If you have a lot of chiseling to do, slide a piece of rubber hose over the chisel. This will lessen the shock to your hand.

When using a chisel for chipping, always wear goggles to protect your eyes. If other men are working close by, see that they are protected from flying chips by erecting a screen or shield to contain the chips.

files

There are a number of different types of files in common use, and each type may range in length from 3 to 18 inches.

Files are graded according to the degree of fineness, and according to whether they have single- or double-cut teeth.

Single-cut files have rows of teeth cut parallel to each other. These teeth are set at an angle of about 65 degrees with the center line. Use single-cut files for sharpening tools, finish filing, and draw filing. They are also the best tools for smoothing the edges of sheet metal and burrs on pipe.

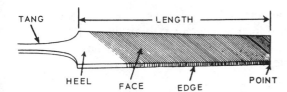

FIGURE 2-21
Parts of a file, without handle.

FIGURE 2-22
File blade shapes in cross section.

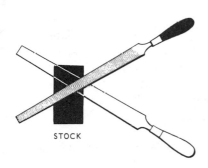

FIGURE 2-23
The proper way to stroke with a file.

Files with crisscrossed rows of teeth are double-cut files. The double cut forms teeth that are diamond-shaped and fast cutting. Use double-cut files for quick removal of metal, and for rough work.

Files are also graded according to the spacing and size of their teeth, or their coarseness and fineness. Some of these grades are pictured in Fig. 2-24. In addition to the grades shown, you may use some dead smooth files, which have very fine teeth, and some rough files with very coarse teeth. The fineness or coarseness of file teeth is also influenced by the length of the file. The length of a file is the distance from the tip to the

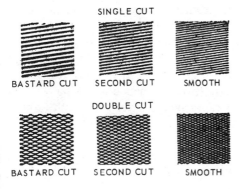

FIGURE 2-24
Closeup view of types of file teeth.

heel, and does not include the tang. When you have a chance, compare the actual size of the teeth of a 6-inch, single-cut smooth file and a 12-inch, single-cut smooth file; you will notice that the 6-inch file has more teeth per inch than the 12-inch file.

shapes

Files come in different shapes. Therefore, in selecting a file for a job, the shape of the finished work must be considered. Some of the cross-sectional shapes are shown in Fig. 2-22.

filing

Using a file is nearly indispensable when one is working with metal. You may be cross filing, draw filing, using a file card, or even polishing metal. Let us examine these operations. (When you have finished using a file, it may be necessary to use an abrasive cloth or paper to finish the product. Whether this is necessary depends on how fine a finish you want on the work.)

CROSS FILING   Cross filing means that the file is moved across the surface of the work in approximately a crosswise direction. For best results, keep your feet spread apart to steady yourself as you file with slow, full-length, steady strokes. The file cuts as you push it; ease up on the return stroke to keep from dulling the teeth. Keep the file clean.

When an exceptionally flat surface is required, file across the entire length of the stock. Then, using the other position, file across the entire length of the stock again. Because the teeth of the file pass over the surface of the stock from two directions, the high spots and low spots are readily visible after you have filed both ways. Continue filing first in one position or direction and then the other until the surface has been filed flat. Test the flatness with a straight edge.

DRAW FILING   Draw filing produces a finer surface finish and usually a flatter surface than cross filing. Small parts are best held in a vise. Hold the file as shown in Fig. 2-26; notice that the cutting stroke is away from you when the handle of the file is held in the right hand. If the handle is held in the left hand, the cutting stroke is toward you. Lift the file away from the surface of the work on the return stroke. When draw filing will no longer improve the surface texture, wrap a piece of abrasive cloth around the file and polish the surface as shown.

USE OF FILE CARD   As you file, the teeth of the file may clog with some of the metal filings and scratch your work. This condition is known as

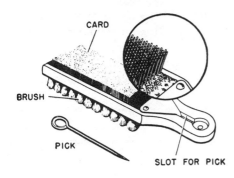

FIGURE 2-25

A file card is used to clean chips from a file.

pinning. You can prevent it by keeping the file teeth clean. Rubbing chalk between the teeth will help prevent pinning, but the best method is to clean the file frequently with a file card or brush. A file card has fine wire bristles. Brush with a pulling motion, holding the card parallel to the rows of teeth.

Always keep the file clean, whether you are filing mild steel or other metals. Use chalk liberally when filing nonferrous metals.

FILING ROUND METAL STOCK   As a file is passed over the surface of round work, its angle with the work automatically changes. A rocking motion of the file results, permitting all the teeth on the file to make contact and cut as they pass over the work's surface. Properly done, the motion tends to keep the file much cleaner and thereby produces better work.

POLISHING A FLAT METAL SURFACE   When polishing a flat metal surface, first draw file the surface as shown. Then finish with abrasive cloth, often called *emery cloth.* Select a grade of cloth suited to the draw filing. If the draw filing was well done, only a fine cloth will be needed to do final polishing.

If your cloth is in a roll, and the job you are polishing is the size that would be held in a vise, tear off a 6- or 8-inch length of the 1- or 2-inch width. If you are using sheets of abrasive cloth, tear off a strip from the long edge of the 8- by 11-inch sheet.

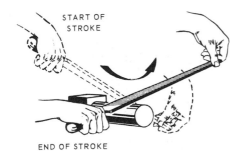

FIGURE 2-26

The proper method for polishing pipe with a file.

Wrap the cloth around the file and hold the file as you would for draw filing. Hold the end of the cloth in place with your thumb. In polishing, apply a thin film of lubricating oil on the surface and use a double stroke with pressure on both the forward and the backward strokes. Note that this is different from the draw filing stroke, in which you cut with the file in only one direction.

When further polishing does not appear to improve the surface, you are ready to use the next finer grade of cloth. Before changing to the finer grade, however, reverse the cloth so that its back is toward the surface being polished.

Work the reversed cloth back and forth in the abrasive-laden oil. Then clean the work thoroughly with solvent before proceeding with the next finer grade of cloth. Careful cleaning between grades helps to insure freedom from scratches.

For the final polish, use a strip of crocus cloth—first use the face and then the back—with plenty of oil. When polishing is complete, again carefully clean the job with a solvent and protect it from rusting with oil or other means.

Another way to polish is with the abrasive cloth wrapped around a block of wood.

POLISHING ROUND METAL STOCK   Round stock may be polished with a strip of abrasive cloth, which is seesawed back and forth over the surface.

Remember that the selection of grades of abrasive cloth, the application of oil, and the cleaning between grades, applies to polishing regardless of how the cloth is held or used.

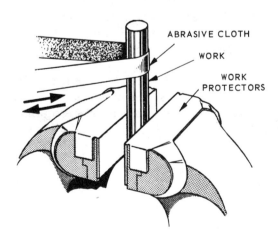

ABRASIVE CLOTH

WORK

WORK
PROTECTORS

FIGURE 2-27

A good way to hold the work in a vise in order to polish the end.

care of files

A new file should be broken in carefully by using it first on brass, bronze, or smooth cast iron. Just a few of the teeth will cut at first, so use a light pressure to prevent tooth breakage. Do not break in a new file by using it first on a narrow surface.

Protect file teeth by hanging your files in a rack when they are not in use, or by replacing them in drawers with wooden partitions. Files should not be allowed to rust—keep them away from moisture. Avoid getting the files oily. Oil causes a file to slide across the work and prevents fast, clean cutting. Files that you keep in your toolbox should be wrapped in paper or cloth to protect their teeth and prevent damage to other tools.

Never use a file for prying or pounding. The tang is soft and bends easily. The body is hard and extremely brittle. Even a slight bend or dropping it can cause a file to snap in two. Do not strike a file against the bench or vise to clean it—use a file card.

safety

Never use a file unless it is equipped with a tightly fitting handle. If you use a file without the handle and it bumps something or jams, the tang may be driven into your hand. To put a handle on a tang, drill a hole in the handle slightly smaller than the tang. Insert the tang end, and then tap the end of the handle to seat it firmly. Make sure you get the handle on straight.

twist drills

Making a hole in a piece of metal is generally a simple operation, but in most cases is an important and precise job. A large number of different tools and machines have ben designed so that holes may be made speedily, economically, and accurately in all kinds of material.

The most common tool for making holes in metal is the twist drill. It consists of a cylindrical piece of steel with spiral grooves, commonly called a *bit*. One end of the bit is pointed, and the other is shaped so that it may be attached to a drilling machine. The grooves, usually called *flutes,* may be cut into the steel cylinder, or may be formed by twisting a flat piece of steel into a cylindrical shape.

The principal parts of a twist drill are the body, the shank, and the point. The dead center of a drill is the sharp edge at the extreme tip end of the drill. It is formed by the intersection of the cone-shaped surfaces of the point and should always be in the exact center of the axis of the drill. The point of the drill should not be confused with the dead center. The point is the entire cone-shaped surface at the end of the drill.

The shank is the part of the drill that fits into the socket, spindle, or chuck of the drill. Several types exist.

A tang is found only on tapered-shank drills. It is designed to fit into a slot in the socket or spindle of a machine. It may bear a portion of the driving torque, but its principal use is to make it easy to remove the drill from the socket or chuck of the driving machine.

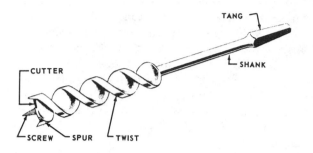

FIGURE 2-28

Parts of an auger bit. It comes in various sizes and makes holes in wood.

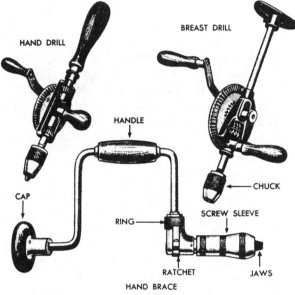

FIGURE 2-29

Types of hand drills. They can be equipped with various types of cutting bits.

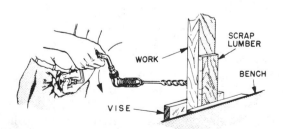

FIGURE 2-30

The proper method for using a hand brace.

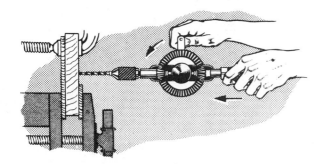

FIGURE 2-31
Using a hand drill.

Twist drills are provided in various sizes. They are sized by letters, numerals, and fractions.

Sets of drills are usually made available according to the way the sizes are stated; that is, "sets of letter drills" or "sets of number drills." However, twist drills of any size (letter, number, or fraction) are available individually if desired.

pipe and tubing cutters and flaring tools

Pipe cutters are used to cut pipe made of steel, brass, copper, wrought iron, and lead. Tube cutters are used to cut tubing made of iron, steel, brass, copper, and aluminum. The essential difference between pipe and tubing is that tubing has considerably thinner walls. Flaring tools are used to make single or double flares in the ends of tubing.

Sizes of hand pipe cutters vary. Examples: The no. 1 pipe cutter has a cutting capacity of ⅛ to 2 inches, and the no. 2 pipe cutter has a capacity of 2 to 4 inches. The pipe cutter has a special alloy-steel cutting wheel and two pressure rollers, which are adjusted and tightened by turning the handle.

Most tube cutters closely resemble pipe cutters, except that they are of lighter construction. A hand-screw-feed tubing cutter of ⅛-inch to 1¼-inch capacity has two rollers with cutouts located off center so that cracked flares may be held in them and cut off without waste of tubing. This cutter is shown at the top of Fig. 2-32. It also has a retractable cutter blade that is adjusted by turning a knob. The other tube cutter shown at the bottom of Figure 2-32 is designed to cut tubing up to and including 1 inch outside diameter. Rotation of the triangular portion of the tube cutter within the tubing will eliminate any burrs.

Flaring tools are used to flare soft copper, brass, or aluminum. The single flaring tool consists of a split die block that has holes for ³⁄₁₆-, ¼-, ⁵⁄₁₆-, ⅜-, ⁷⁄₁₆-, and ½-inch outside diameter (o.d.) tubing, a clamp to lock the tube in the die block, and a yoke that slips over the die

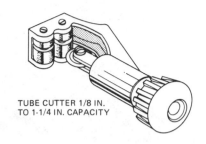

TUBE CUTTER 1/8 IN.
TO 1-1/4 IN. CAPACITY

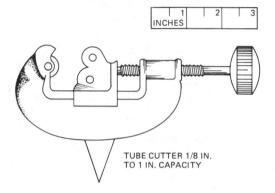

TUBE CUTTER 1/8 IN.
TO 1 IN. CAPACITY

FIGURE 2-32
Tubing cutters.

block. It has a compressor screw and a cone that forms a 45-degree flare or a bell shape on the end of the tube. The screw has a T-handle. A double flaring tool has the additional feature of adapters that turn in the edge of the tube before a regular 45-degree double flare is made. It consists of a die block with holes for $\frac{3}{16}$-, $\frac{1}{4}$-, $\frac{5}{16}$-, $\frac{3}{8}$-, and $\frac{1}{2}$-inch tubing, a yoke with a screw and a flaring cone, plus five adapters for different sizes of tubing, all carried in a metal case.

WOOD-CUTTING SAWS

Wood-cutting handsaws come in a variety of types. Two useful ones for the plumber are the so-called crosscut saw and the ripsaw. The crosscut is for cutting boards across or at an angle to the grain of the wood; the rip is for cutting with the grain, or lengthwise.

Saws come with varying numbers of teeth per inch. The more teeth the saw has, the smoother the cut will be. For general use, one with 8 teeth per inch will serve well. A 26-inch blade is a preferred size. To tell quality, look for a springy, well-finished blade.

saw precautions

A saw that is not being used should be hung up or stored in a toolbox. A toolbox designed for holding saws has notches that hold them on edge, teeth up. Storing saws loose in a toolbox may allow the saw teeth to become dulled or bent by coming into contact with other tools.

Before using a saw, be sure that there are no nails or other edge-destroying objects in the line of the cut. When sawing out a strip of waste, do not break out the strip by twisting the saw blade. Doing so dulls the saw and may spring or break the blade.

Be sure that the saw will go through the full stroke without striking the floor or some object. If the work cannot be raised high enough to obtain full clearance for the saw, you must carefully limit the length of each stroke.

using a hand saw

To saw across the grain of the stock, use the crosscut saw; to saw with the grain, use a ripsaw. Study the teeth in both kinds of saws (Fig. 2-33) so that you can readily identify the saw that you need.

TOP VIEW OF RIP TEETH TOP VIEW OF CROSSCUT TEETH

TEETH OF RIP SAW TEETH OF CROSSCUT SAW

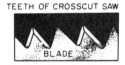

FIGURE 2-33

Closeup view of rip and crosscut saw teeth. The ripsaw is used to cut boards lengthwise, the crosscut to cut across the grain.

Place the board on a saw horse or some other suitable object. Hold the saw in the right hand (or the left hand, if you are left-handed) and extend the first finger along the handle. Grasp the board and take a position so that an imaginary line passing lengthwise along the right forearm will be at an angle of approximately 45 degrees with the face of the board. Be sure that the side of the saw is plumb or at right angles with the board face. Place the heel of the saw on the mark. Keep the saw in line with the forearm and pull it toward you to start the cut.

To begin, take short, light strokes, gradually increasing the strokes to the full length of the saw. Do not force or jerk the saw. Such procedure only makes sawing more difficult. The arm that does the sawing should

FIGURE 2-34

The proper position of the body when sawing.

swing clear of your body so that the handle of the saw operates at your side rather than in front of you.

Use one hand to operate the saw. You may be tempted to use both hands at times, but if your saw is sharp, one hand will serve you better. The weight of the saw is sufficient to make it cut. Should the saw stick or bind, it may be because the saw is dull and is poorly set, or because the wood has too much moisture in it, or because you have forced the saw and thus have caused it to leave the straight line.

Keep your eye on the line rather than on the saw while sawing. Watching the line lets you see instantly any tendency to leave it. A slight twist of the handle, and taking short strokes while sawing will bring the saw back. Blow away the sawdust frequently so that you can see the line.

Final strokes of the cut should be taken slowly. Hold the waste piece in your other hand so that the stock will not split when you take the last stroke.

Short boards may be placed on one sawhorse to be sawed. Place long boards on two sawhorses, but do not saw so that your weight falls between them, or your saw will bind. Place long boards so that your weight is directly on one end of the board over one sawhorse, while the other end of the board rests on the other sawhorse.

Short pieces of stock are more easily cut when they are held in a vise. When you rip short stock, it is important that you keep the saw from sticking, so it may be necessary to take a squatting position. The saw can then take an upward direction and thus work more easily. When you rip long boards, it will probably be necessary to use a wedge in the saw kerf to prevent binding (Fig. 2-35).

FIGURE 2-35
To prevent binding, a wedge may be inserted in the cut (kerf) when sawing.

SCREWDRIVERS

A screwdriver is one of the most basic of basic hand tools. It is also the most frequently abused of all hand tools. It is designed for one function only—to drive and remove screws. A screwdriver should not be used as a pry bar, a scraper, a chisel, or a punch.

standard

There are three main parts to a standard screwdriver. The portion you grip is called the *handle,* the steel portion extending from the handle is the *shank,* and the end that fits into the screw is called the *blade.*

The steel shank is designed to withstand considerable twisting force in proportion to its size, and the tip of the blade is hardened to keep it from wearing.

Standard screwdrivers are classified by size, according to the combined length of the shank and blade. The most common sizes range in length from 2½ to 12 inches. There are many smaller and some larger for special purposes. The diameter of the shank, and the width and thickness of the blade are generally proportionate to the length, but again there are special screwdrivers with long thin shanks, short thick shanks, and extra wide or extra narrow blades.

Screwdriver handles may be wood, plastic, or metal. When metal handles are used, there is usually a wooden hand grip placed on each side of the handle. In some types of wood- or plastic-handled screwdrivers the shank extends through the handle, whereas in others the shank enters the handle only a short way and is pinned to the handle. For heavy work, special types of screwdrivers are made with a square shank. They are designed in this way so that they may be gripped with a wrench, but these types are the only kind on which a wrench should be used.

When you use a screwdriver, it is important to select the proper size so that the blade fits the screw slot properly. Proper fit prevents burring the slot and reduces the force required to hold the driver in the slot. Keep the shank perpendicular to the screw head.

Another type of screwdriver that is essential for your toolbox is the Phillips. The head of a Phillips-type screw has a four-way slot into which the screwdriver fits. Three standard-sized Phillips screwdrivers handle a wide range of screw sizes. Their ability to hold helps to prevent damaging the slots or the work surrounding the screw. It is a poor practice to try to use a standard screwdriver on a Phillips screw, because both the tool and screw slot will be damaged. Use the same technique for driving and removing screws as for standard screwdrivers.

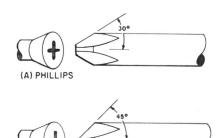

(A) PHILLIPS

(B) REED AND PRINCE

FIGURE 2-36
Tips of two types of screwdrivers. The Phillips is far more common than the Reed and Prince.

PLIERS

Pliers are made in many styles and sizes and are used to perform many different operations, but mainly to hold and grip small objects in situations where it may be inconvenient or impossible to use hands.

Combination pliers are handy for holding or bending flat or round stock. *Long-nosed* pliers are less rugged, and break easily if you use them on heavy jobs. Long-nosed pliers (commonly called needle-nose pliers) are especially useful for holding small objects in tight places and for making delicate adjustments. The *round-nosed* kind are handy when you need to crimp sheet metal or form a loop in a wire.

Slip-joint pliers have straight, serrated (grooved) jaws, and the screw or pivot with which the jaws are fastened together may be moved to either of two positions in order to grasp small- or large-sized objects better.

To spread the jaws of slip-joint pliers, first spread the ends of the handles apart as far as possible. The slip-joint, or pivot, will move to the open position. To close, again spread the handles as far as possible, then push the joint back into the closed position.

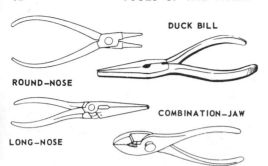

DUCK BILL

ROUND–NOSE

COMBINATION–JAW

LONG–NOSE

FIGURE 2-37
Types of pliers.

wrench (vise-grip) pliers

Vise-grip pliers can be used for holding objects regardless of their shape. A screw adjustment in one of the handles makes them suitable for several different sizes. The jaws of vise-grips may have serrations or may be clamp-type. The clamp-type jaws are generally wide and smooth and are used primarily for working with sheet metal.

Vise-grips have an advantage over other types of pliers in that you can clamp them on an object and they will stay, leaving your hands free for other work.

A craftsman uses this tool in a number of ways. It may be used as a clamp, a speed wrench, and a portable vise, as well as in many other applications where a locking, plier-type jaw may be employed. These pliers can be adjusted to various jaw openings by turning the knurled adjusting screw at the end of the handle. Vise-grips can be clamped and locked in position by pulling the lever toward the handle.

Vise-grip pliers should be used with care, since the teeth in the jaws tend to damage the object on which they are clamped. Do not use them on nuts, bolts, tube fittings, or other objects that must be reused.

water-pump pliers

Water-pump pliers were originally designed for tightening or removing water-pump packing nuts. They were excellent for this job, because they have a jaw adjustable to seven different positions. Water-pump pliers are easily identified by their size, jaw teeth, and adjustable slip joint. The inner surface of the jaws consists of a series of coarse teeth formed by deep grooves, a surface adapted to grasping cylindrical objects.

channel-lock pliers

Channel-lock pliers are another version of water-pump pliers easily identified by the extra-long handles, which make them a very powerful gripping

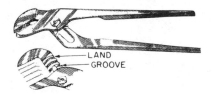

LAND
GROOVE

FIGURE 2-38

Channel-lock pliers. They are similar in appearance to water-pump pliers.

tool. Their shape is approximately the same as that of the pliers just described, but the jaw opening adjustment is different. Channel-lock pliers have grooves on one jaw and lands on the other. The adjustment is effected by changing the position of the grooves and lands. Consequently, these pliers are less likely to slip from the adjustment setting when gripping an object. They should be used only where it is impossible to use a more adapted wrench or holding device. Many nuts and bolts and surrounding parts have been damaged by improper use of channel-lock pliers.

VISES AND CLAMPS

Vises are used for holding work when it is being sawed, drilled, shaped, sharpened, riveted, or glued. Clamps are used for holding work that cannot be satisfactorily held in a vise because of its shape and size, or in the absence of a vise. Clamps are generally used for light work.

A machinist's bench vise is a large steel vise with rough jaws that prevent work from slipping. Most of these vises have a swivel base with jaws that can be rotated. A similar light-duty model is equipped with a cutoff. These vises are usually bolt-mounted onto a bench.

The bench and pipe vise has jaws for holding pipe from ¾ inch to 3 inches in diameter. The maximum main jaw opening is usually 5 inches, with a jaw width of 4 to 5 inches. The base can be swiveled to any position and locked. These vises are equipped with an anvil and are bolted onto a workbench.

The pipe vise is specifically designed to hold round stock or pipe. The one shown in Fig. 2-39 has a capacity of 1 to 3 inches. One jaw is hinged so that the work can be positioned and then the jaw brought down and locked. This vise is also used on a bench. Some pipe vises are designed to use a section of chain to hold down the work. Chain pipe vises range in size from ⅛- to 2½-inch pipe capacity up to ½- to 8-inch capacity.

A C-clamp is shaped like the letter C. It consists of a steel frame threaded to receive an operating screw with a swivel head. It is made for light, medium, and heavy service in a variety of sizes.

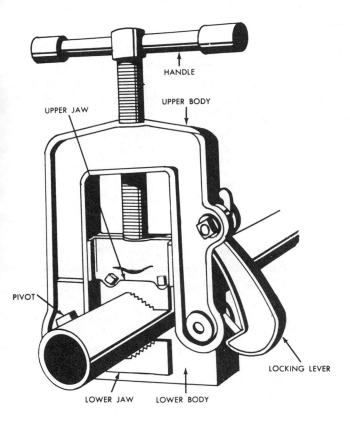

HANDLE

UPPER BODY

UPPER JAW

PIVOT

LOWER JAW   LOWER BODY

LOCKING LEVER

FIGURE 2-39
A pipe vise.

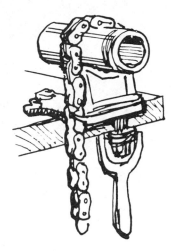

FIGURE 2-40
A chain vise is good for
handling large pipe.

care

Keep vises clean at all times. They should be cleaned and wiped with
light oil after using. Never strike a vise with a heavy object and never
hold large work in a small vise, since these practices will cause the jaws to
become sprung or otherwise damage the vise. Keep jaws in good condition and oil the screws and the slide frequently. Never oil the swivel base
or swivel jaw joint; its holding power will be lessened. When the vise is not
in use, bring the jaws lightly together or leave a very small gap. (Heat may
cause the movable jaw of a tightly closed vise to break because of expansion of the metal.) Leave the handle in a vertical position.

Threads of C-clamps must be kept clean and free from rust. The
swivel head must also be clean, smooth, and grit free. If the swivel head
becomes damaged, replace it as follows: Pry open the crimped portion of
the head and remove the head from the ball end of the screw. Then replace
with a new head and crimp.

safety precautions

1. When closing the jaw of a vise or clamp, avoid getting your hands or body between the jaws or between one jaw and the work.
2. When holding heavy work in a vise, place a block of wood under the work as a prop to prevent it from sliding down and falling on your foot.
3. Do not open the jaws of a vise beyond their capacity, as the movable jaw will drop off, perhaps hurting you.

SABER SAW

The saber saw is a power-driven jigsaw that will let you make curved and straight cuts in wood and light metal. Most are light-duty machines and are not designed for extremely fast cutting. A prime use is cutting notches in framing.

There are several different blades designed to operate in the saber saw, and they are easily interchangeable. For fast cutting of wood, a blade with coarse teeth may be used. A blade with fine teeth is designed for cutting metal.

The best way to learn how to handle this type of tool is to use it. Before trying to do a job, clamp down a piece of scrap plywood and draw some curved as well as straight lines to follow, and practice cutting. You will develop your own way of gripping the tool, and this will be affected to some degree by the particular tool you are using. On some tools, for example, you will find guiding easier if you apply some downward pressure on the tool as you move it forward. If you are not firm with your grip, the tool will tend to vibrate excessively, roughening the cut. Do not force the cutting faster than the design of the blade allows, or you will break it.

RULES AND TAPES

There are many different types of measuring tools. If very accurate measurements are required, a caliper is used. On the other hand, if accuracy is not extremely critical, the common rule or tape suffices for most measurements.

A steel rule will serve well. This rule is usually 6 or 12 inches in length, although other lengths are available. Steel rules may be flexible, but the thinner the rule, the easier it is to measure accurately, because the divisions of marks are closer to the work.

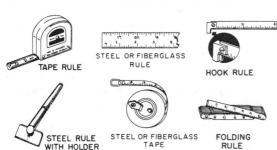

TAPE RULE

STEEL OR FIBERGLASS RULE

HOOK RULE

STEEL RULE WITH HOLDER

STEEL OR FIBERGLASS TAPE

FOLDING RULE

FIGURE 2-41

Types of rules. Tapes and folding rules are most valuable to the plumber.

Generally a rule has four sets of graduations, one on each edge of each side. The longest lines represent the inch marks. On one edge, each inch is divided into eight equal spaces, so each space represents ⅛ in. The other edge of this side is divided into sixteenths. The ¼-in. and ½-in. marks are commonly made longer than the smaller division marks to make counting easier, but the graduations are not usually numbered individually, as they are sufficiently far apart to be counted without difficulty. The opposite side is similarly divided into 32 and 64 spaces per inch, and it is common to number every fourth division for easier reading.

There are many variations of the common rule. Sometimes graduations are on one side only; sometimes a set of graduations is added across one end for measuring in narrow spaces; sometimes only the first inch is divided into sixty-fourths, with the remaining inches divided into thirty-seconds and sixteenths.

A metal or wood folding rule may be used for measuring purposes. Folding rules are usually two to six feet long. The folding rules cannot be relied on for extremely accurate measurements, because a certain amount of play develops at the joints after the ruler has been used for a while.

Steel tapes are made in lengths of six feet and longer. The shorter lengths are frequently made with a curved cross section so that they are flexible enough to roll up, but remain rigid when extended. Long, flat tapes require support over their full length when you are measuring, or the natural sag will cause an error in reading.

The flexible-rigid tapes are usually contained in metal cases into which they wind themselves when a button is pressed, or into which they can be easily pushed. A hook is provided at one end to hook over the object being measured so that one man can handle it without assistance. On some models, the outside of the case can be used as one end of the tape when inside dimensions are measured.

measuring procedures

To take a measurement with a common rule, hold the rule with its edge on the surface of the object being measured. Doing so will eliminate errors that may result because of the thickness of the rule. Read the measurement at the graduation that coincides with the distance to be measured, and state it as being so many inches and fractions of an inch. Always reduce fractions to their lowest terms. For example, $6/8$ inch would be called $3/4$ inch. A hook or eye at the end of a tape or rule is normally part of the first measured inch.

outside pipe diameters

To measure the outside diameter of a pipe, it is best to use some kind of rigid rule. A folding wooden or steel rule is good enough. As shown in Fig. 2-42, line up the end of the rule with one side of the pipe, using your thumb as a stop. Then with the one end held in place with your thumb, swing the rule through an arc and take the maximum reading at the other side of the pipe. For most practical purposes, the measurement obtained by using this method is satisfactory. It is necessary that you know how to take this measurement, as the outside diameter of pipe is sometimes the only dimension given on pipe specifications.

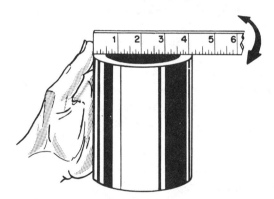

FIGURE 2-42
Measuring outside diameter of pipe.

To measure the inside diameter of a pipe with a rule, hold the rule so that one corner of it just rests on the inside of one side of the pipe.

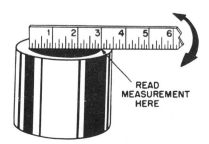

READ
MEASUREMENT
HERE

FIGURE 2-43
Measuring inside diameter of pipe.

Then, with one end held in place, swing the rule through an arc and read the diameter across the maximum inside distance. You will obtain an approximate inside measurement.

pipe circumferences

To measure the circumference of a pipe, a flexible rule that will conform to the cylindrical shape of the pipe must be used. A tape rule or a steel tape is good. When measuring, make sure that the tape has been wrapped squarely around the axis of the pipe to insure that the reading will not be more than the actual circumference. This is extremely important when large diameter pipe is measured.

Hold the rule or tape as shown in Fig. 2-44. Take the reading, using the 2-inch graduation, for example, as the reference point. In this case the correct reading is found by subtracting two inches from the actual reading. In this way the first two inches of the tape, serving as a handle, will enable you to hold the tape securely.

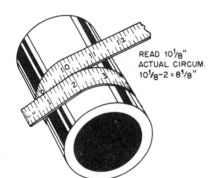

READ 10⅛"
ACTUAL CIRCUM.
10⅛−2 = 8⅛"

FIGURE 2-44
Measuring circumference of pipe.

inside dimensions

To take an inside measurement, such as the inside of a box (or studs), a folding rule that incorporates a 6- or 7-inch sliding extension is one of the best measuring tools to use. To take the inside measurement, first unfold the folding rule to the approximate dimension. Then extend the end of the rule and read the length that it extends, adding the length of the extension to the length on the main body of the rule.

To measure an inside dimension using a tape rule, extend the rule between the surfaces, take a reading at the point on the scale where the rule enters the case, and add two inches. The two inches is usually the width of the case. The total is the inside dimension being taken.

To measure the thickness of stock through a hole with a hook rule, insert the rule through the hole, hold the hook against one face of the stock, and read the thickness at the other face.

outside dimensions

To measure an outside dimension using a tape rule, hook the rule over the edge of the stock. Pull the tape out until it projects far enough from the case to permit measuring the required distance. The hook at the end of the rule is designed so that it will locate the end of the rule at the surface from which the measurement is being taken. When a length measurement is taken, the tape is held parallel to the edge. For measuring widths, the tape should be at right angles to the lengthwise edge. Read the dimension of the rule exactly at the edge of the piece that is being measured.

It may not always be possible to hook the end of the tape over the edge of stock being measured. If so, it may be necessary to butt the end of the tape against another surface or to hold the rule at a starting point from which a measurement is to be taken.

distance measurements

Tapes are generally used for making long measurements. Secure the hook end of the tape. Hold the tape reel in your hand and allow it to unwind while you walk in the direction in which the measurement is to be taken. Stretch the tape with sufficient tension to overcome sagging; at the same time make sure that the tape is parallel to an edge or to the surface being measured. Read the distance.

care

Rules and tapes should be handled carefully and kept lightly oiled to prevent rust. Never allow the edges of measuring devices to become nicked by striking them with hard objects. It is preferable to keep them in a wooden box when not in use.

To avoid kinking tapes, pull them straight out from their cases— do not bend them backward. With the windup type, always turn the crank clockwise; turning it backward will kink or break the tape. With the spring-wind type, guide the tape by hand. If it is allowed to snap back, it may be kinked, twisted, or otherwise damaged. Do not use the hook as a stop. Slow down as you reach the end.

SQUARES, PLUMB BOB, AND LEVELS

Squares are primarily used for testing and checking trueness of an angle or for laying out lines on materials. Most squares have a rule marked on their edge. As a result they may also be used for measuring. For the plumber, a carpenter's square is most useful.

carpenter's square

The size of a carpenter's steel square is usually 12×8 inches, 24×16 inches, or 24×18 inches. The flat sides of the blade and the tongue are graduated in inches and fractions of an inch. (The square also contains information that helps to simplify or eliminate the need for computations in many woodworking tasks.) The most common uses for the square are laying out and squaring up large patterns, and for testing the flatness and squareness of large surfaces. Squaring is accomplished by placing the square at right angles to adjacent surfaces and observing if light shows between the work and the square.

plumb bob

A plumb bob is a pointed, tapered, brass or bronze weight that is suspended from a cord for determining the vertical or plumb line to or from a point on the ground. Common weights for plumb bobs are 6, 8, 10, 12, 14, 16, 18, and 24 oz.

A plumb bob is a precision instrument and must be cared for as such. If the tip becomes bent, the cord from which the bob is suspended will not occupy the true plumb line over the point indicated by the tip. A plumb bob usually has a detachable tip; if the tip should become damaged, it can be replaced without replacing the entire instrument.

The plumb bob is used to determine true verticality when vertical uprights and corner posts of framework are erected. Surveyors use it for transferring and lining up points.

To locate a point that is exactly below a particular point in space, secure the plumb bob string to the upper point. When the plumb stops swinging, the point indicated at B in Fig. 2-45 will be exactly below A.

To plumb a structural member or a pipe, secure the plumb line A so that you can look at both the line and the piece behind the line. Then, by sighting, line up the member or pipe with the plumb line. If it is impossible to sight directly, it may be necessary to secure the plumb line at some

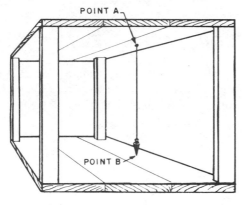

FIGURE 2-45
A plumb bob is used to determine true verticality.

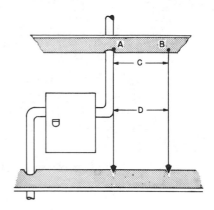

FIGURE 2-46
A method for finding the distance between two true verticals using a plumb bob.

point such as B, and then measure the offset from the line to the piece at two places so that, for example, C and D are equal. If the distances C and D are not equal, adjust the structural member or conduit until they are.

levels

Levels are tools designed to prove whether a plane or surface is a true horizontal or true vertical.

The level is a simple instrument consisting of a liquid, such as alcohol or chloroform, partially filling a glass vial(s) or tube(s) so that a bubble remains. The tube is mounted in a frame, which may be aluminum, wood, or iron. Levels are equipped with one, two, or more tubes. One tube is built in the frame at right angles to another. On the outside of the tube are two sets of graduation lines separated by a space. Leveling is accomplished when the air bubble is centered between the graduation lines.

Levels must be checked for accuracy. Place the level on a true horizontal surface and note where the bubble is. Reverse the level end for end. The bubble should also be centered.

Do not drop or handle a level roughly. To prevent damage, store it in a rack or other suitable place when not in use.

METAL-CUTTING OPERATIONS

Many hand tools and power tools have been designed for the specific purpose of cutting metals quickly and accurately. This section will describe

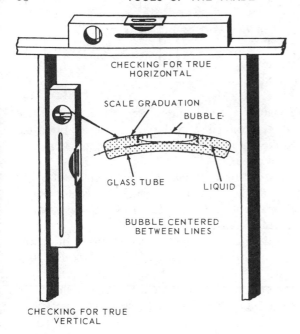

CHECKING FOR TRUE
HORIZONTAL

SCALE GRADUATION

BUBBLE

GLASS TUBE   LIQUID

BUBBLE CENTERED
BETWEEN LINES

CHECKING FOR TRUE
VERTICAL

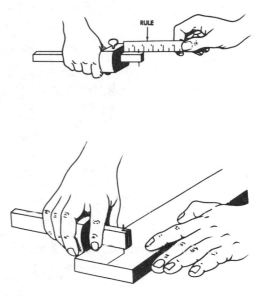

RULE

FIGURE 2-47

A level is good for establishing level and true of short distances.

FIGURE 2-48

A scratch gauge is used for marking boards.

some metal-cutting operations that can be performed with chisels, drills, taps, dies, reamers, and pipe and tubing cutters.

metal cutting with chisels

When struck with a heavy hammer, a cold chisel is capable of cutting metal. With chisel and hammer, you can cut wires, bars, rods, and other shapes of metal and also cut off the heads of rivets and bolts.

cutting wire or round stock

Mark off a guide line on the stock and place the work on the top face of an anvil or other suitable working surface. Place the cutting edge of the chisel on the mark in a vertical position and lightly strike the chisel with a hammer. Check the chisel mark for accuracy. Continue to strike the chisel until the cut is made. The last few blows of the hammer should be made lightly to avoid damage to the anvil, the supporting surface, or the chisel.

Heavy stock is cut in the same manner, except that the cut is made halfway through the stock; then the work is turned over, and the cut is finished from the opposite side.

cutting off a rivet or bolt head

Hold the work in a heavy vise, or secure it some other way so that the work will not move. Hold the cold chisel with one face of the bevel flat on the surface of the job. Strike the head of the chisel with the hammer as you loosely hold and guide the chisel.

To cut off a rivet head with a cape chisel, select a chisel of about the same size as the diameter of the rivet. Cut through the center of the rivet head, holding one face of the bevel flat on the surface of the job, and then sever the center of the head from the shank or body.

To cut off a rivet head with a side cutting chisel, place the chisel nearly flat on the surface of the work with its single bevel upwards. Drive the cutting edge under the edge of the rivet head just as you would if you were using a cold chisel. The cutting edge of the chisel has a slight radius that will tend to prevent the corners from cutting undesirable grooves in the surface of the work.

To remove a rivet head when there is not room enough to swing a hammer with sufficient force to cut the rivet, first drill a hole about the size of the body of the rivet in and almost to the bottom of the rivet head. Then cut off the head with a cold chisel.

cutting holes in metal

In drilling any metal, there are several general steps to be followed. First, mark the exact location of the hole. Second, secure the work properly. Then use the correct cutting speed and appropriate cutting oil or other coolant. Finally, apply pressure on the drill properly. It is assumed that you have selected the correct drill size.

The exact location of the hole must be marked with a center punch. The punch mark forms a seat for the drill point, insuring accuracy. Without the mark, the drill may have a tendency to "walk off" before it begins to cut into the metal.

holding the work

Most work is held for drilling by some mechanical means such as a vise or clamps. If it is not well secured, the work may rotate at high speed or fly loose, and become a high-speed projectile.

When one is drilling in small pieces with a hand-held drill, it is best to hold the work in a vise so that the axis of the drill is horizontal.

This position provides better control and tends to insure a hole that is straight.

Just as lumber is the carpenter's basic material in doing his job, pipe is the basic material to the plumber. An understanding of what is available and how to work with it is essential to proficiency in the trade. Following is a roundup of the kinds of pipe there are, the fittings or connectors used with the pipe, and techniques on how to work with pipe.

PLASTIC PIPE

This is the newest pipe of all. Nearly two-thirds of all plastic piping used for DWV systems is made of ABS (acrylonitrile-butadiene-styrene). It is a tough, high-impact material that has been used in millions of DWV installations as well as in building sewers and sewer mains. ABS is also suited to conduit and pressure pipe application.

Polyethylene (PE) is flexible and can be obtained in several grades and densities. It is used for water service, gas distribution and service, irrigation, and chemical waste systems.

A member of the rigid family of plastics, PVC (polyvinyl chloride) has a broad range of applications, which include water mains and water service, sewer mains and building sewers, DWV, gas distribution and gas service, irrigation, electrical and communications conduit and industrial process piping, and many others.

A cousin of PVC, CPVC (chlorinated polyvinyl chloride) has been chemically formulated to withstand higher temperature than other plastics. It is nominally rated at 180°F at 100 psig, with a substantial safety margin built in.

3

Pipe: Selecting and Working with It

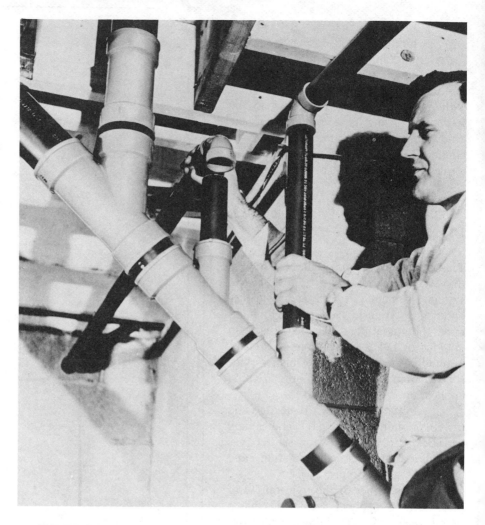

FIGURE 3-1
If allowed by the local building code, plastic pipe works well in a drainage system.

CPVC is used for domestic hot and cold water piping systems as well as for industrial process piping applications.

Styrene rubber (SR) is plastic impregnated with particles of rubber. It is mainly used in septic tank leaching fields, subsoil dewatering systems, and other nonpressure applications.

As a general rule, plastics are limited only by temperature-pressure capabilities. The higher the operating pressure of the system, the less will be the temperature capability of plastic, and vice versa. For example, in a drainage system relatively high temperatures can be handled, because the system is not under pressure. However, the rating of most plastic pressure piping is based on 73°F laboratory test temperature.

The greatest use has been in water systems. These include rural and municipal water mains, water service piping, and, more recently as a result of the development of CPVC, hot and cold distribution piping within the building. Plastic is claimed to have superior flow characteristics and resistance to corrosion and scale buildup.

as sewer pipe

Leaks in sewers allow ground water to enter the sewer during rainy periods, thereby overloading sewerage treatment facilities. These same leaks can allow raw sewage to contaminate underground water sources. Plastic piping is said to have superior impact strength, high joint integrity, and resistance to corrosion, making it useful in this application.

Of all the markets into which plastics have ventured, the greatest acceptance has been in sanitary drainage (DWV) piping. There are at least five million DWV installations in the United States. Although plastic DWV has been used primarily in residential construction, it is gaining rapid acceptance in commercial, industrial, and institutional buildings as well. Plastic piping can also be used for industrial process piping, irrigation, and sprinkler systems.

support of plastic pipe

Plastic piping installed above ground is hung with the same devices used for metal piping, except that care must be taken so that the hanger does not cut or abrade the pipe. Rigid horizontal piping three inches or less in diameter should be supported at intervals of no more than three feet. Piping four inches and above in diameter should be supported at intervals of no more than five feet. Vertical piping may be supported through the use of standard pipe clamps, so long as they do not cut or abrade the piping.

Rigid plastics have a certain degree of flexibility, so they must be buried so that the trench bottom supports the pipe throughout its entire length. Do not use blocks to attain proper grade. A select backfill having a particle size preferably of not more than ¾-inch diameter is placed and tamped to at least six inches above the top of the pipe. Thereafter, any available backfill can be used.

working plastic pipe

Plastic pipe is the easiest of all pipe to work, but because of different characteristics of each type of plastic, there are a number of methods for joining the pipe to itself as well as to other materials. They are:

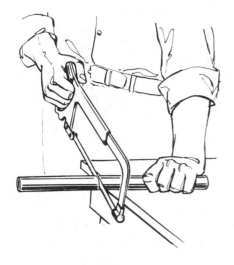

FIGURE 3-2

Plastic pipe is easy to work with. It can be cut easily with a hacksaw. Burrs should then be removed with a knife.

1. SOLVENT CEMENTING   Some people who see this process for the first time think of it as "gluing" the pipe together. Actually, through the use of the solvent cement, the installer is chemically welding the pipe to the fitting.

2. INSERT FITTINGS   Flexible plastics such as PE can be joined through the use of insert fittings that are pushed into the pipe and secured with stainless steel clamps.

3. ELASTOMERIC SEALS   The use of bell-end piping with elastomeric seals is becoming an increasingly popular method of joining plastic piping, particularly in the larger diameters and in underground installations. This joint goes together quickly.

4. THREADED FITTINGS   Some plastic pipe can be threaded. Thread cutters especially designed for the job are available.

5. HEAT FUSION   Certain plastics, such as PE, are so chemical-resistant that there is not a solvent strong enough to fuse them. Therefore, heat fusion is used in certain applications to join these materials.

6. TRANSITION FITTINGS   There are a number of fittings specially designed to connect plastics to other piping materials. Plastics can be connected, by means of these fittings, to any other materials including cast iron, steel, copper, clay, asbestos cement, and concrete.

GALVANIZED STEEL PIPE

Older water supply systems commonly use galvanized steel pipe. It is good for withstanding the pressures of water hammer and high water pressure.

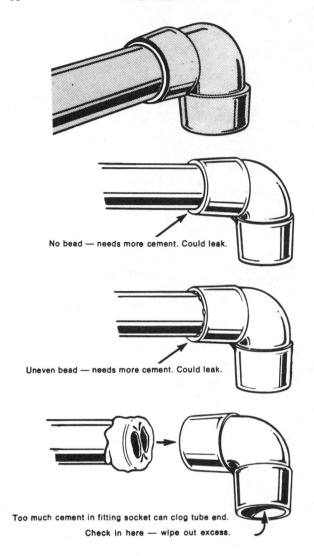

No bead — needs more cement. Could leak.

Uneven bead — needs more cement. Could leak.

Too much cement in fitting socket can clog tube end.

Check in here — wipe out excess.

FIGURE 3-3

Examples of correct and incorrect joints made with plastic pipe. The top joint—with a continuous bead of cement around the joint—is correct; the others are incorrect.

It rarely leaks, because the threaded connections tend to seal each other.

Galvanized steel pipe has a couple of drawbacks. It is threaded, and sometimes the threading must be done by the plumber. Although it can be done by hand, it is a long, tiresome job. Much better is to do it with an electric vise. You stick the pipe in the vise, turn it on, and the pipe rotates and is threaded automatically. Another disadvantage of galvanized steel is that it tends to corrode and scale with certain types of water. Since it has been used so widely, you will find that many fixtures and appliances come with threaded pipe tappings. However, galvanized pipe should not be used underground.

FIGURE 3-4
One method of joining
plastic pipe is using
a cement that is
applied with a dabber
appropriate to the
size of the pipe.

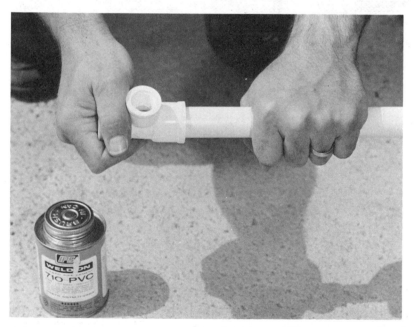

FIGURE 3-5
The plastic fitting is
secured to the pipe by
pushing it over the
end of the pipe, giving
it a quarter turn, and
allowing it to cure.
Curing time varies
according to
temperature and
other factors.

FIGURE 3-6
Existing metal pipe setup.

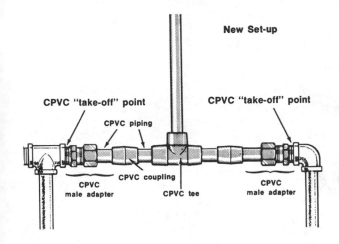

New Set-up

CPVC "take-off" point   CPVC "take-off" point

CPVC piping

CPVC coupling

CPVC male adapter   CPVC tee   CPVC male adapter

FIGURE 3-7
Fittings are available to let you use plastic pipe on an existing metal pipe setup, such as the one shown in Figure 3-6.

Galvanized pipe is also used for the DWV system, but it is not very popular. For this, 1½-, 2-, and 3-inch diameters are used, along with special drainage fittings. Malleable fittings should be used with water supply pipe.

It should be noted that galvanized water supply pipe can be simple to work with if a system can be figured out so that all sections used are bought prethreaded. Usual sizes used are ⅜, ½, ¾, and 1 inch. These dimensions refer to the nominal inside diameter. The outside diameter usually measures about ¼ inch more. Pipe is commonly available in 21-foot lengths with both ends threaded and one coupling loosely screwed to one end.

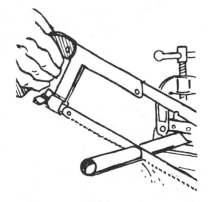

FIGURE 3-8
Before cutting metal pipe, secure it in a pipe vise.

FIGURE 3-9
A hacksaw can be used to cut heavy walled pipe.

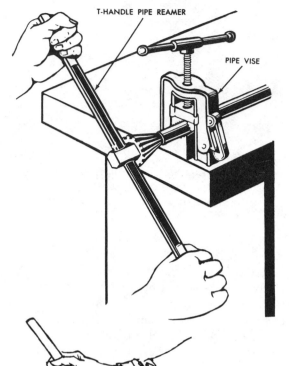

T-HANDLE PIPE REAMER

PIPE VISE

FIGURE 3-10

After cutting metal pipe, ream the end to remove burrs.

FIGURE 3-11

Use a thread cutter with the appropriate size die. Place it at a right angle to the pipe and rotate it slowly until the die grabs hold.

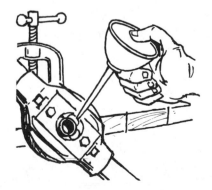

FIGURE 3-12

Give the thread cutter a half turn; then back it off a quarter turn and apply cutting oil. Repeat this operation until the threads are formed.

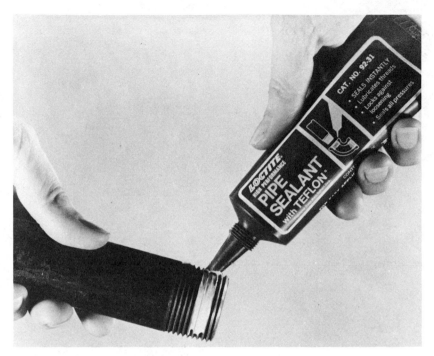

FIGURE 3-13
After the thread is cut, apply joint compound, Teflon tape or other sealer to the metal pipe before inserting it into the fitting.

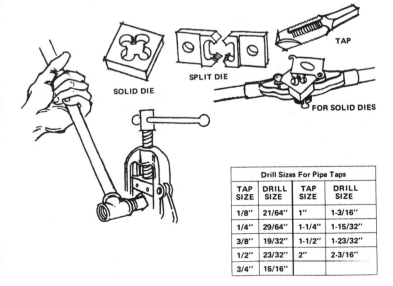

SOLID DIE

SPLIT DIE

TAP

FOR SOLID DIES

Drill Sizes For Pipe Taps			
TAP SIZE	DRILL SIZE	TAP SIZE	DRILL SIZE
1/8"	21/64"	1"	1-3/16"
1/4"	29/64"	1-1/4"	1-15/32"
3/8"	19/32"	1-1/2"	1-23/32"
1/2"	23/32"	2"	2-3/16"
3/4"	15/16"		

FIGURE 3-14
A variety of dies and taps for working with metal pipe is available.

T

90° L
(Reducing)

T
(Reducing)

90° L

REDUCER

PLUG

BUSHING

PIPE CAP

45° L

45° L (Street)

GROUND-JOINT
UNION

DROP-EAR L

EXTENSION
PIECE

90° L
(Street)

COUPLING

CROSS

FLOOR FLANGE

FIGURE 3-15
Examples of fittings available for use with galvanized pipe.

FIGURE 3-16
When tightening or loosening a fitting, use two wrenches— one as a vise to hold the pipe and the other to turn the fitting.

CAST IRON PIPE

Most plumbers agree that the best piping for the DWV system (i.e., stacks, building drains, and sewer pipes; galvanized pipe is often used with it for branch, vent, and secondary stacks) is cast iron. It is strong and has great resistance to corrosion. The standard method for joining cast iron pipe is by stuffing the joint or hub where a plain pipe end fits with oakum, then pouring in hot lead. This is not the easiest work in the world, and it can be dangerous if water is accidentally introduced to the hot lead (it can explode). In recent years, however, the industry has come up with so-called no-hub pipe, as described by the Cast Iron Soil Pipe Institute. Instead of the so-called bell and spigot joint, the pipes and fittings are plain-ended. To join pipe, the ends are slipped into a neoprene shield and then the shield is tightened with stainless steel clamps. The joint is leak-

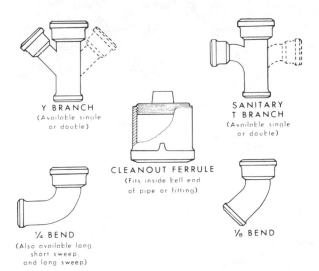

Y BRANCH
(Available single
or double)

CLEANOUT FERRULE
(Fits inside bell end
of pipe or fitting)

SANITARY
T BRANCH
(Available single
or double)

¼ BEND
(Also available long
short sweep,
and long sweep)

⅛ BEND

FIGURE 3-17
Examples of fittings available for use with cast iron pipe.

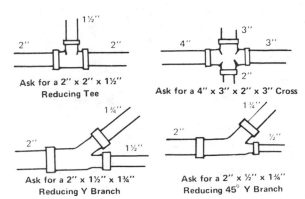

Ask for a 2″ x 2″ x 1½″
Reducing Tee

Ask for a 4″ x 3″ x 2″ x 3″ Cross

Ask for a 2″ x 1½″ x 1¾″
Reducing Y Branch

Ask for a 2″ x ½″ x 1¼″
Reducing 45° Y Branch

FIGURE 3-18
The method for ordering reducing fittings.

proof but flexible. If you make a mistake, you can take it apart in a couple of minutes and refit it. The only tool required is a 60-lb torque wrench. You tighten first one side of the clamp and then the other (like tightening studs on a car tire) until a 48-lb torque is achieved.

FIGURE 3-19
To cut heavy cast iron pipe, first score around it with a cold chisel; then break it off with a hammer.

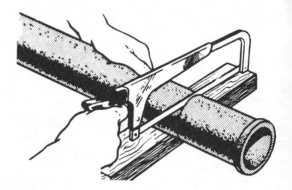

FIGURE 3-20
To cut light cast iron soil pipe, first score it all around it, following a chalkline; then make a shallow cut all around with a hacksaw.

FIGURE 3-21
Then, with the pipe on a board as shown, tap it with a hammer until the piece breaks off.

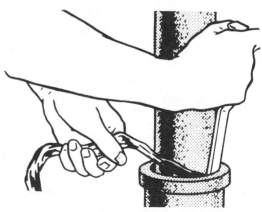

FIGURE 3-22
To join cast iron pipe with bell and spigot ends, first stick the spigot end of the pipe into the hub of the lower pipe. With a yarning iron, wedge oakum into the joint.

FIGURE 3-23
Then carefully pour lead into the joint until the hub is full. Make sure there is no moisture present when using hot lead and that the ladle is heated before using. A small explosion could result if these precautions are not taken.

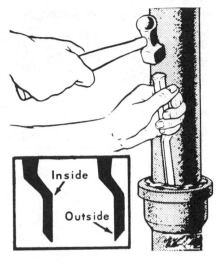

FIGURE 3-24
Finally, when the lead is cool, pack it down with an outside caulking iron, then an inside caulking iron.

FIGURE 3-25
If you have to pour lead on a horizontal joint, use a device called an asbestos joint runner to hold the lead in while it cools.

This no-hub system can be used in an existing drain line. You simply cut out parts of the pipe not wanted and install the new sections with the shields and clamps. Fittings are also designed with the no-hub feature.

Cast iron pipe is not easy to cut. It requires a hacksaw, cold chisel, and hammer, although one company does sell a special chain-type cutter that is good for cast iron pipe. If you are working with the bell and spigot system, you must allow (as with any pipe-fitting arrangement) for the distance that the pipe will go into the fitting when cutting that pipe. With the hubless system you do not have to worry about makeup—just cut the lengths long enough so they seat firmly in the shield.

Hubless pipe comes in 5- and 10-foot lengths and 2-, 3-, and 4-inch diameters (inside diameters). You can use it above or below

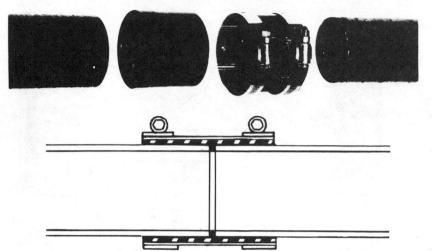

FIGURE 3-26
Hubless cast iron pipe is easy to use. Cut pipe ends are inserted into a neoprene gasket until they seat against an inner shoulder in the gasket. Then the stainless steel clamp is tightened with a torque wrench to uniform tightness.

ground, in any drainage situation. If the branch drain is less than 2 inches, the pipe is made with 1½ inch galvanized steel pipe and drainage-type threaded fittings used in the no-hub cappings.

VITRIFIED CLAY PIPE

Vitrified clay pipe, or glazed pipe, is a material that used to be common underground drainage line usage. It is acid resistant (or should be), making it a good choice for laying in soil. If the soil is not compact, but gives poor support, the pipe must be supported by boards underneath, so that it cannot bend and crack and rupture joints.

Vitrified clay pipes come with a socket or bell-shaped end that accepts the spigot end when the pipe is joined. The laying lengths—that is, the pipe without the fittings—are available in lengths of 2 feet, 2 feet 6 inches, and 3 feet, and in diameters of from 3 to 42 inches.

Standard fittings used with this type pipe include Y branches, double Y branches, short and long elbows, and V branches.

BRASS PIPE

Brass pipe is the most expensive you can buy. It is commonly used for water supply systems where water is particularly corrosive. Like galvanized steel, nails will not puncture it, and high pressures do not affect it. It also has the advantage of exceptionally smooth walls, which allow water to flow very freely.

Also like galvanized pipe, you sometimes have to thread it yourself, and doing so is not easy. When you are working with brass pipe and fittings, it is a good idea to use a strap wrench, because this protects its shiny finish.

Brass may be connected with galvanized steel pipe, but it is a good idea to use a connector, or dielectric coupling, between the two to prevent harmful metallic action. You can join copper and plastic pipe and fittings to brass without connectors.

COPPER PIPE

There are two principal categories of copper pipe. The first is commonly referred to as *plumbing tube* and includes types K, L, M, and DWV. The second is ACR tube, for air conditioning and refrigeration field service.

Types K, L, and M, and DWV are designated by the American Society for Testing and Materials (ASTM) standard sizes, with the outside diameters always ⅛ inch larger than the standard size. Each type represents a series of sizes with different wall thicknesses. The inside diameters depend on tube size and wall thickness. All four types are available in so-called "drawn" temper in straight lengths ordinarily 20 feet long. Types K and L are available also in annealed temper in 20-foot straight lengths or in coils in sizes from ¼ inch to 2 inches. Drawn tube is often referred to as "hard" tube, and annealed tube as "soft." Drawn tube is also available in a "bending temper," and may be specified for jobs requiring bending. Tube of a hard temper can be joined by soldering or brazing, capillary fittings, or by welding. Tube in the bending and soft tempers can be joined the same way and also by the use of flare-type compression fittings.

The most common method of joining copper tube is by soldering with capillary fittings. Soldered joints are used for water lines, and for sanitary drainage. Brazed joints, with capillary fittings, are used where greater strength is required or where temperatures during use are as high as 350°F. Brazing is preferred, and often required, for joints in refrigeration piping. Mechanical joints are used frequently for underground tubing, for joints that may have to be disconnected from time to time. Copper tube may also be joined by butt-welding without the use of fittings.

Cast bronze pressure fittings are available in all standard tube sizes and in a wide variety of types for plumbing and heating. They can be either soldered or brazed, although brazing cast fittings requires care.

Wrought copper pressure fittings are also available over a wide range of sizes and types. These, too, can be joined by either soldering or brazing and are preferred where brazing is the joining method. Otherwise, the choice between cast and wrought fittings is not critical.

Flared-tube fittings provide metal-to-metal contact similar to ground joint unions; both can be easily taken apart and reassembled. They are especially useful where water cannot be removed from the tube, mak-

STEP ①

SCREW THE CUTTING WHEEL
LIGHTLY AGAINST THE TUBING

FIGURE 3-27
Using a tubing cutter.

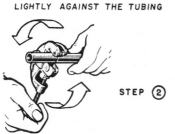

STEP ②

ROTATE THE CUTTER KEEPING A SLIGHT
PRESSURE AGAINST THE CUTTING WHEEL
WITH THE SCREW ADJUSTMENT.

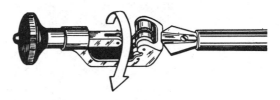

FIGURE 3-28
Some cutters have a pointed end for reaming pipe.

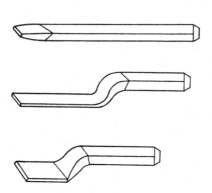

FIGURE 3-29
Tools needed for cast iron: cold chisel (top), yarning iron (middle), and caulking iron (bottom). (Both inside and outside caulking irons are available.)

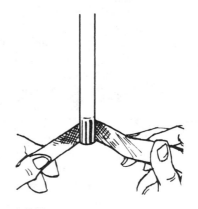

FIGURE 3-30
After cutting copper pipe, use emery cloth to clean both the end of the tubing and the inside of the fitting it will go into.

FIGURE 3-31
After cleaning, apply
noncorrosive flux to the end
of the copper pipe and the
inside of the fitting. Rotate
the pipe inside the fitting to
spread the flux evenly.

FIGURE 3-32
Holding solder to the work,
apply heat to the work (not
the solder) so capillary
action draws solder into the
joint (hence the name
capillary fittings).

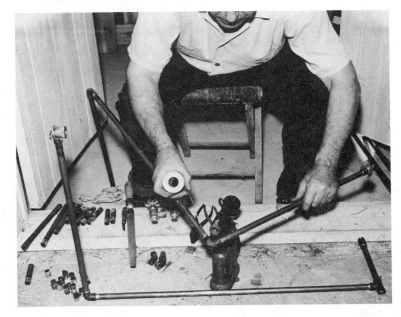

FIGURE 3-33
It's best to assemble
fittings on a handy
workbench, or even the
floor, before installing in
the system.

ing soldering difficult. Flared joints may be required where a fire hazard
exists or a torch is not allowed. Also, soldering under wet conditions can
be very difficult, and flared joints are preferred in such cases.

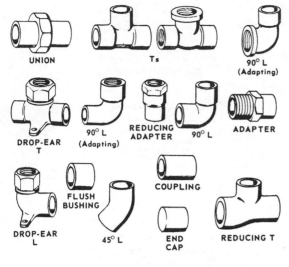

UNION

Ts

90° L
(Adapting)

DROP-EAR
T

90° L
(Adapting)

REDUCING
ADAPTER

90° L

ADAPTER

DROP-EAR
L

FLUSH
BUSHING

COUPLING

45° L

END
CAP

REDUCING T

FIGURE 3-34

Examples of various solder fittings used with copper pipe.

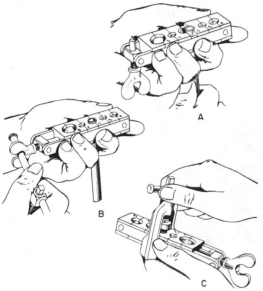

A

B

C

FIGURE 3-35

Many valves and fittings are of the flare type. Here the pipe is flared before connection with a die block: (A) Pipe is inserted; (B) Clamp is tightened; (C) Flare is made with a punch.

FIGURE 3-36

Flared pipe and fitting.

MEASURING PIPE FOR CUTTING

Before cutting pipe it is, of course, necessary to measure it. Following are the methods to use for various kinds of pipe.

Galvanized Steel. Here you must first allow for the distance that the pipe will go into the fittings. The best way to do this is with the face-to-face method. Measure the exact distance between the face of one fitting and the other, then tack on the length or distance that the pipe will go into the fittings. For example, if the distance between one fitting and the other is three feet and the pipe goes into each fitting one-half inch, you need to cut a 3-foot, 1-inch length. The larger the pipe, the further it will go into the fittings. For example, 1-inch pipe will go in ⅝ inch; 1½-inch pipe, ⅝ inch; 2-inch pipe, ¾ inch.

Plastic. Here, you can use the face-to-face method as above.

Cast Iron. The hubless variety of this pipe is easy to measure, since both ends are plain, fitting into a neoprene washer. If you are using the old hub and spigot pipe, you must allow for the pipe's going into the fitting or hub end. You should allow 2½ inches for 3-inch pipe and 3 inches for 4-inch pipe. If the pipe is less than five feet long (cast iron comes in 5- and 10-foot lengths), use double-hub pipe—it has a hub on each end. When you cut it, you have a hub on each end; otherwise, a plain, cut-off end would be useless.

Copper. For rigid tubing, you can use the face-to-face method of measuring, allowing for the distance that the tube will go into the fittings. If you are using flexible tubing with flare fittings, use the face-to-face method and allow about ¾₁₆ inch for the flare fitting.

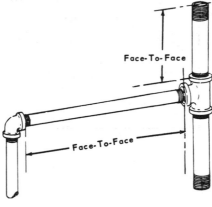

Face-To-Face

Face-To-Face

MEASURING PIPE RUNS

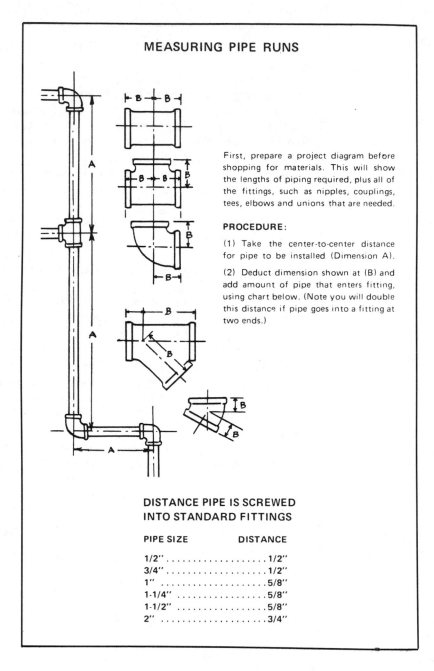

First, prepare a project diagram before shopping for materials. This will show the lengths of piping required, plus all of the fittings, such as nipples, couplings, tees, elbows and unions that are needed.

PROCEDURE:

(1) Take the center-to-center distance for pipe to be installed (Dimension A).

(2) Deduct dimension shown at (B) and add amount of pipe that enters fitting, using chart below. (Note you will double this distance if pipe goes into a fitting at two ends.)

DISTANCE PIPE IS SCREWED INTO STANDARD FITTINGS

PIPE SIZE	DISTANCE
1/2"	1/2"
3/4"	1/2"
1"	5/8"
1-1/4"	5/8"
1-1/2"	5/8"
2"	3/4"

FITTINGS

Every type of pipe is available with an assortment of fittings, which are the connectors used to join the pipe. With fittings you can make pipes branch out, go around turns, and connect to other kinds of pipe. The following are the names of ones that are commonly used and what they are.

1. *Nipples* range in size from 1⅜ inches long to around 12 inches and are used to join pipe where only short sections are needed. They come threaded and unthreaded, depending on the type of pipe involved.
2. *Couplings* connect pipes of similar size that will not have to be disconnected.
3. *Unions* connect pipe that is expected to be taken apart at some future date.
4. *Dielectric couplings* connect threaded pipes of dissimilar metals. They are designed to prevent undesirable metallic reaction.
5. *Reducing couplings* connect pipes of different sizes that will not have to be disconnected.
6. *Elbows* join pipes at 45-degree or 90-degree angles.
7. *Adapters* let you join a small pipe to a large pipe.
8. *Reducing T's* join three pipes at a 90-degree angle.
9. *Crosses* join four pipes.
10. *Bushings* let you join pipe and fittings of dissimilar size.
11. *Plugs* are used to seal off a female end of a fitting.
12. *Caps* are used to close off one end of a pipe.
13. *Y branches* join two pipes plus a third pipe running out at a 45-degree angle.

The term *fittings* is also used to describe faucets and valves, but its largest application is for joining pipe.

4

The job of bringing the drainage or water supply lines—any pipe in the plumbing system—through walls and floors to fixtures where the connections are made is called *roughing in;* in essence, it refers to the pipe that will be out of sight in the structure.

As such, it has to be especially carefully installed, so that leaks of air or water do not occur later, whether from vibration, poor joints, settling of the house or ground, or radical temperature changes. It will, of course, be necessary to cut a way for the piping through a house, and this must be done carefully so that structural members, such as joists and studs, are not weakened; if necessary, these can be braced for strength. You also want to do the minimum amount of cutting. One cardinal rule is not to join piping in holes that are too small; house movement can rupture the pipe.

In installing the piping for an entire house, the drainage system normally starts with the installation of the house drain, and then the drainage piping is installed; the water pipe follows the same general route and can be run in the same spaces as the drainage pipe.

In order that fixtures be properly connected, and located in relation to the rough plumbing, manufacturers furnish plans containing roughing-in dimensions. Typical roughing-in sketches are shown in Figure 4-1.

CUTTING THE WAY

When cutting through house framing, find the size of the studs and joists and

Roughing In

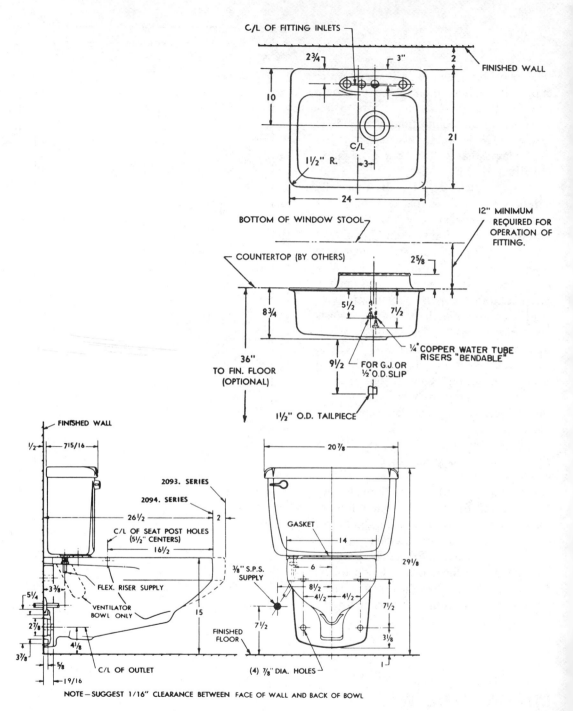

FIGURE 4-1

Typical roughing-in sketches.

refer to the table below to find space needed. For example, a 2 × 4 stud wall has about 3¾ inches clearance inside. It will not take 2-inch cast iron pipe (with hubs), but it will accept 3-inch copper or plastic pipe with fittings. A 2 × 6 joist has about 5¾ inches of clearance; it will take a 2-inch cast iron pipe; a 2 × 8 has about 7¾ inches of clearance—it will take up to 4 inches of cast iron.

PIPE SIZE	CAST-IRON		PLASTIC & COPPER	
	Pipe	Fittings	Pipe	Fittings
1-1/2''	–	–	1-3/4''	2-1/8''
2''	4''	4''	–	–
3''	–	–	3-3/4''	3-5/8''
4''	6-1/4''	6-1/4''	–	–

FIGURE 4-2

Measuring stud depth (X).

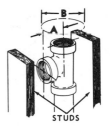

FIGURE 4-3

Pipe must have room inside walls both in terms of depth (A) and width (B).

It should be noted that these dimensions are based on current dimensions of finished lumber. A trend in the lumber industry is to make the finished lumber sizes smaller and smaller, so a good procedure is to measure the boards first to determine exactly how much clearance you have.

Plastic and copper pipes are easily cut to any length, and fittings can be located as desired; you can run 1½-inch pipe (without fittings) through a space as small as 1¾ inches, or 3-inch pipe through a space only 3¼ inches wide. Fittings can then be installed where extra space is available. Another advantage of plastic or copper pipe is that no turning space is required.

NOT ENOUGH SPACE

You will run into situations where partition walls do not have enough space for pipes. The solution is to build the studs out a bit to provide the space, and then resurface with new wall material. One way to do this with 2 × 4 studs is shown in Fig. 4-4.

If an outside wall is too thin for pipes and you cannot break into

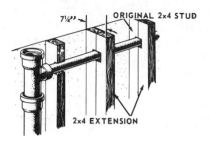

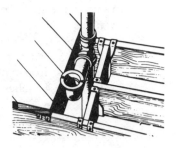

FIGURE 4-4
The easy way to thicken a wall for pipe fitting is to insert 2 × 4 extensions and resurface.

FIGURE 4-5
Toilet drains should run between joists or below them whenever possible.

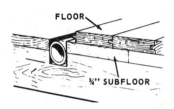

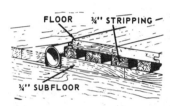

FIGURE 4-6
Sometimes you can run a pipe through a floor rather than cutting through a joist.

it, because either the plaster or plasterboard is against brick or other hard-surface outer wall material, you may have to add 2 × 6 or 2 × 8 studs to fatten the wall and increase space inside.

NOTCHING STUDS

It will also be necessary, in running pipes throughout a house, to notch boards so the pipe can fit in. You must take care here, because if the studs are load-bearing (support a floor and wall), then what you are essentially doing, as mentioned earlier, is weakening it. To make sure you do it correctly, follow these tips

1. Do not cut (notch) more than one-third to one-half the depth of the stud without reinforcing it with metal strap.

2. Do not notch deeper than two-thirds the stud even if you plan on using a metal strap.

3. The upper half of a stud may be notched to one-half its depth only if only two studs in a row are being notched and there are two unnotched studs left.

FIGURE 4-7
You shouldn't cut (or notch) a stud or joist deeper than ⅓ of its depth without reinforcing it with a metal strap. Drilling holes is the preferred method of running pipe through framing members.

It should be emphasized that notching is not the preferred method, and many plumbers—and codes—do not use or allow it. Drilling holes through the framing members is the preferred method.

5

An important part of plumbing work is installing fixtures, particularly the sink, tub, and toilet. Following is information on what is available and tips on installation.

Before you install new fixtures in an existing bath, it is well to think a bit. A new bath job may range from just taking out the fixtures and installing new ones to ripping out everything to the studs and starting from scratch. If it is a very old bathroom, probably the best thing is to go down to the studs. If there is only one bath in the house, you will have to go slowly. Plan your work so that you do not inconvenience anyone. Take the fixtures out only when you are ready to get the new ones in right away. Of course, if the work is new, you need not worry about any of the above.

REMOVING FIXTURES

There are certain things that you must do when removing fixtures. The first is to turn off the water supply. Usually, as explained earlier, there are valves for turning off the basin and toilet. Basin valves (there are two) are under the sink; the toilet valve is under the water closet. Tub valves (two) are in the cellar beneath the tub, or they may be in a panel in a wall behind the tub; in a slab house the valves may be in the utility room. If you cannot find the appropriate valves, you can turn off the main valve in the basement, turning off all water, hot and cold, in the entire house.

When you disconnect any fixture, pay careful attention to how it is connected. It will make the job of putting the new one in easier.

Fixtures

removing kitchen sinks

To remove a kitchen sink, you will have to loosen clips that hold the sink to the countertop. The faucet has two thin supply pipes leading from it that hook up with water pipes coming out of the wall; a basin wrench is handy for removing nuts that are up close to the back of the sink.

The basin also is tied into the waste line. Here, there is a slip nut (you pull it up with your fingers) that holds the drain pipe in place in the trap. When all nuts are loose, you can just lift the sink up and out of the way.

removing toilets

First, flush the bowl. Use a pot to remove all water from the bowl (or as much as you can); sponge out the rest. One type of bowl has a flush elbow, which is connected to the tank. Unscrew with a wrench and take the bowl and tank out separately. New types of toilets have the tank and bowl in one piece. In either case, toilets are connected to the floor flange with two or four bolts that must be removed before you can lift the toilet out of the way.

removing tubs

If the tub is the old-fashioned type with decorative "feet," the piping is exposed, and it is a simple matter to disconnect the tub. Just use a wrench on the exposed nuts.

Another type of tub is built into the wall (Fig. 5-1). To remove it, you have to open up the walls first to reach the connection of the tub to the water and waste lines. If the walls are plasterboard, just rip it off with a pry bar, first removing the escutcheons (the decorative pieces below the faucet handles). If the walls are plaster, use a sledgehammer. A hammer is useful in either case.

For removing any tub from the waste line, check downstairs in the cellar (if you can) to see if the waste line from the tub can be unscrewed from the waste line proper. If you can, just loosen the nut and lift straight up on the tub; the attached waste line will come up and out. It may be that the waste line on the tub is connected to the main line by a slip nut instead of a normal nut. If this is the case, just push the slip nut up out of the way with your fingers.

If you cannot disconnect the tub waste line from the waste line proper, the only recourse is to disconnect it at the drain flange from inside the tub. Doing so may be difficult, because it is usually corroded. At

FIGURE 5-1
Some tubs, such as the one shown, are built into the wall.

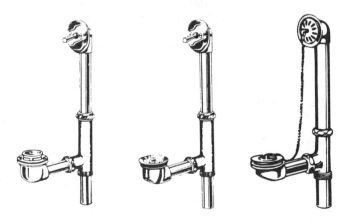

FIGURE 5-2
Three types of tub drains: pop-up
(left); trip lever (center);
connected waste and overflow
with chain and rubber stopper
(right). The first two drains come
in 1 ½″ diameter, the connected
waste and overflow in 1 ⅜″ and
1 ½″ diameters.

any rate, fit the handles of a large wrench into holes in the drain, and then slip a screwdriver between the handles for leverage and turn.

Incidentally, if piping is ancient (lead water pipes), or very old (galvanized iron pipe), or just old (brass pipes), it ought to be replaced with copper pipe. All the pipes that are going to end up covered and in-

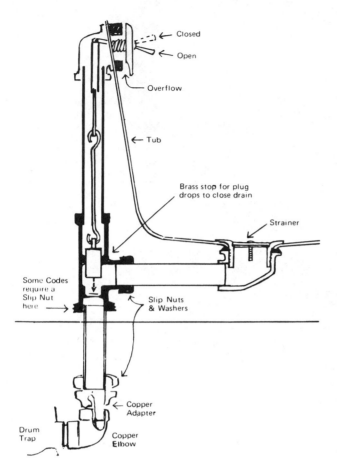

Closed

Open

Overflow

Tub

Brass stop for plug
drops to close drain

Strainer

Some Codes
require a
Slip Nut
here →

Slip Nuts
& Washers

Drum
Trap

Copper
Adapter

Copper
Elbow

FIGURE 5-3

A typical pipe setup for a tub. The trap and drain at the bottom sometimes are used when the tub is not accessible from the basement. Full installation instructions for the bath drain come with the unit.

accessible must be in A-1 condition. You do not want them to develop problems, requiring that the wall be opened up for repairs.

TUBS

selection

Bathtubs are usually made of one of three materials: cast iron with a porcelain enamel coating, steel with porcelain, or molded glass fiber reinforced with a polyester resin; these last tubs are commonly known as fiberglass tubs.

Sizes and shapes vary. Commonly, bathtubs are rectangular, usually four to six feet long (commonly five feet), with heights ranging from

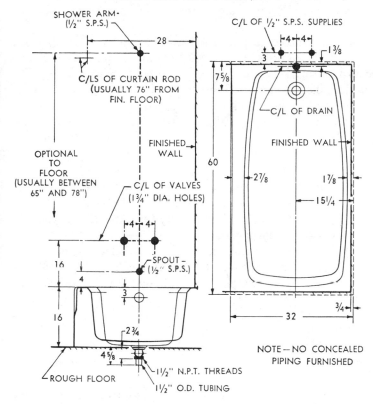

2607.109 - LEFT HAND OUTLET (SHOWN)
2605.103 - RIGHT HAND OUTLET (REVERSE DIMENSIONS)

SHOWER ARM·
(1/2" S.P.S.)

C/L OF 1/2" S.P.S. SUPPLIES

28

C/LS OF CURTAIN ROD
(USUALLY 76" FROM
FIN. FLOOR)

C/L OF DRAIN

FINISHED WALL

FINISHED
WALL

OPTIONAL
TO
FLOOR
(USUALLY BETWEEN
65" AND 78")

C/L OF VALVES
(1¾" DIA. HOLES)

SPOUT –
(1/2" S.P.S.)

ROUGH FLOOR

1½" N.P.T. THREADS
1½" O.D. TUBING

NOTE — NO CONCEALED
PIPING FURNISHED

FIGURE 5-4

*Roughing-in dimensions
are provided by the
manufacturer.*

12 to 16 inches from floor to rim top. The actual water depth of the tub is, of course, from the bottom of the tub to the overflow drain.

Cast iron is the heaviest (up to 500 pounds) of the tubs and is more durable and less susceptible to damage than other types. Formed steel tubs are light (about 100 pounds) and less expensive than the cast iron. Many plumbers use them for upper-story installations because of their manageability and because they do not usually require that the floor be reinforced. Incidentally, some steel tubs are made of two sections welded together, but a seam shows and many people do not like this.

Steel tubs are more apt to make noise than cast-iron tubs. Some companies offer them, therefore, with an undercoating that helps deaden sound.

The fiberglass tub is a relative newcomer to the plumbing scene. The surface is especially smooth, and easy to keep clean, but it does scratch relatively easily. Some companies make kits that you can use to remove scratches. Coming into more popular use lately is the one-piece fiberglass tub-shower unit. An advantage of this type of unit is that it eliminates the need for making a joint where the tub meets the wall; this joint often breaks open, allowing water to penetrate into walls behind the tub.

Fiberglass tub-shower combination units come in many different sizes and are molded in various shapes, but the most common is the five-foot-long rectangular one.

In position, the tub stands on the floor, but the back rim also rests on a ledger strip nailed to the studs. You can make this ledger strip out of any wood available, from a 1 × 3 to a 2 × 6.

installation

First, measure the tub from the underside of the lip to the floor to determine the exact height. It should not be too low, so add an extra ⅛ inch and be certain that it is level. If the floor is very far out of level, you will have to take this condition into account.

Use your level to check floor levelness. If the floor is more than ¼ inch out of level from the back wall to the face of the tub, put the tub in place and then shim it up temporarily with undercourse shingles or pieces of wood.

Use a straightedge board cut to length to level the tub end to end at the back wall and then front to back on each end. Do not be surprised if you cannot get both ends level. Large metal castings are often distorted. If one end happens to be out of level, it should pitch toward the front, the way the water runs.

After you are satisfied that the tub is level, mark the studs where the top of the tub is. Slide the tub away from the wall. Measure the thickness of the back edge of the tub. Allow this amount from the lines on the studs to the top of the ledger and nail the ledger solidly to the studs. Now move the tub back in place and shim the front edge. It should now be level and rest solidly on the ledger and the floor and shims.

If you have easy access to the tub connections from the outside (i.e., if they are open to an unfinished area), a tub is easy to connect. Indeed, always try to trim the tub (put on drain pipes, faucets, etc.) first and work to the trap. If you do not have this advantage of easy access, it will be rather difficult to make the connections.

After the tub is shimmed up, mark the wall where the center of the overflow and drain holes fall. You need three measurements to locate the exact center of each hole (i.e., the distance from two walls and the height off the floor). Nail the shims in place so that the tub can be taken away and replaced without the necessity of leveling it all over again.

Remove the tub. Make whatever alterations are necessary to the floor framing to make room for the drain pipes (one from the overflow, one from the bottom of the tub). Assemble the drain pipes on the tub and

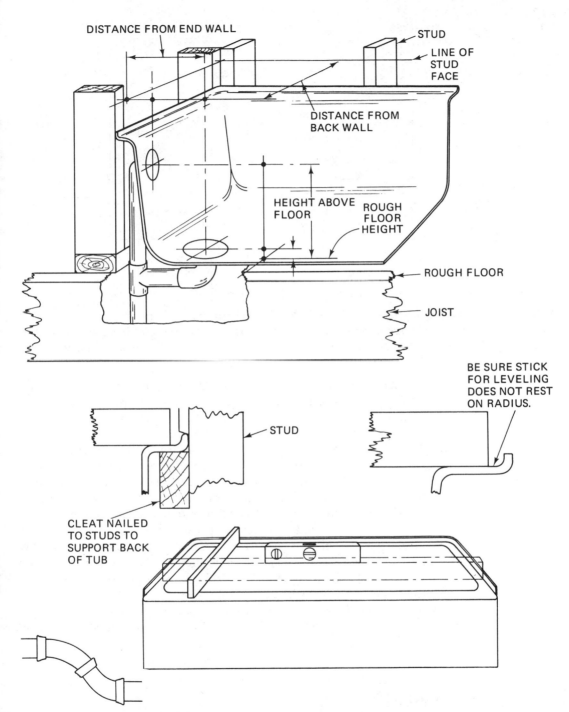

FIGURE 5-5

When installing the tub as shown, it is important to make certain it is level. Use a straightedge board and a level to make sure the tub is level from end to end and from front to back.

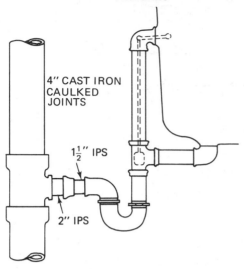

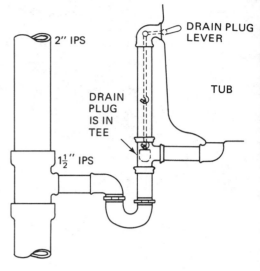

FIGURE 5-6

Arrangement of piping when connecting the drain to a 4" cast iron soil pipe.

FIGURE 5-7

Arrangement of piping when connecting the drain to a 2" galvanized iron pipe.

screw them up tight, and then remove them as an assembly. Hold the assembly in approximate place on the waste pipe according to your measurements to get an idea of where the trap has to be. A number of ways to hook it up are possible. See Figs. 5-6 and 5-7 for possibilities.

The brass tub trap has a joint in it so that the "U" part can swing 180 degrees or more. This gives a great flexibility for alignment. Screw the trap directly on the waste pipe. Then set the whole thing up and make up all the joints tight; adjust the trap to align the waste flange (lip) and overflow flange according to the marks and measurements you took earlier.

Use tape to hold the washers in place on the two flanges temporarily. Then put the tub in place. If you did everything properly, the two flanges will line up with the two holes in the tub. Pack plenty of stainless putty around the drain fitting and screw it in, using the handles of a large pair of pliers and a screwdriver to turn it.

If there is access to the back and bottom of the tub, all the fittings can be made up after the tub is in place. The trap can be used to make the final adjustment.

TOILETS

Toilets are made in two basic shapes—elongated and round. Colors vary (colored fixtures are more expensive than white), but toilets are usually made of vitreous china. As durable as this material is, it is subject to scratching and chipping and should be handled with care on the job.

APPROXIMATE DIMENSIONS FOR TOILETS			
	TANK		EXTENSION OF FIXTURE INTO
	HEIGHT (inches)	WIDTH (inches)	ROOM (inches)
One-piece toilet	18½ to 25	26¾ to 29¼	26¾ to 29¼
Close-coupled tank and bowl	28½ to 30⅞	20⅝ to 22¼	27½ to 31⅜
Wall-hung toilet	27 to 29½	21 to 22¼	26 to 27½ (concealed tank, 22)
Wall-hung tank	32 to 38	17¾ to 22	26½ to 29½
Corner toilet	28¾	19¼	31

There are three different flushing actions commonly used in toilets for residences. These result from bowl design and, although they have been available for years, few individuals are familiar with their differences or the advantages of one over the other. Consequently, price has frequently been the determining factor in bowl selection, resulting later in dissatisfaction.

bowl designs

WASHDOWN   The washdown bowl discharges into a trapway at the front; it is most easily recognized by a characteristic bulge on the front. It has a much smaller exposed water surface inside the bowl, with a large flat exposed china surface at the front of the bowl interior. Since this area is not protected by water, it is subject to fouling, contamination, and staining. The trapway in the washdown bowl is not round, and its interior is frequently irregular in shape because of the exterior design and the method of manufacture. Characteristically, the washdown bowl does not flush as well or as quietly as other bowls.

This type of bowl is no longer accepted by many municipal code authorities, and many manufacturers have deleted it from their manufacturing schedule.

REVERSE TRAP   The reverse trap bowl discharges into a trapway at the rear of the bowl. Most manufacturers' models have a large exposed water surface, thereby reducing fouling and staining of the bowl interior. The trapway is generally round, providing a more efficient flushing action.

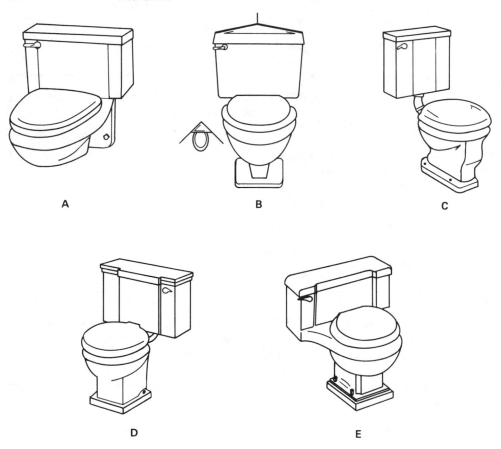

A   B   C

D   E

FIGURE 5-8

Types of toilets: (A) Wall-hung—complete wall-hung toilets make it possible to clean the floor under the toilet. (B) Corner—The corner toilet is a space saver; note the triangular tank. (C) Two-piece with wall-hung tank. (D) Close-coupled tank and bowl—the tank, a separate unit, is attached to the bowl. (E) One-piece—One-piece toilets are neat in appearance and easy to clean, but are more expensive than two-piece models.

SIPHON JET   The siphon jet is similar to the reverse trap in that the trapway also discharges to the rear of the bowl. All models must have a large exposed water surface, leaving less interior china surface exposed to fouling or contamination. The trapway must be large, and it is engineered to be as round as possible for the most efficient flushing action.

The federal government has established minimum trapway diameters and exposed water surface areas for each of the bowl designs. The

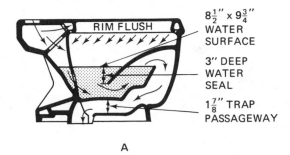

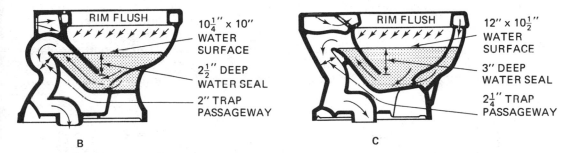

FIGURE 5-9
Flushing actions of toilets: (A) Washdown; (B) Reverse trap; (C) Siphon jet.

basic differences between the siphon jet and the reverse trap bowls are in these areas, both minimums being larger in the siphon jet to assure better flushing action and less bowl contamination.

installing a toilet

There is usually either a 3-inch copper waste pipe or a 4-inch lead pipe coming up through the floor that is 12 inches (to the center of pipe) from the finished wall. On a ceramic tile floor a brass toilet flange is installed directly on the floor; otherwise, you must use a marble slab under it. Different toilet flanges are made for use with lead or copper pipe, and you need one to match the kind of waste pipe there is. If there is copper, the flange you use may be either straight or offset. The offset flange is used to get more or less clearance from the wall if the waste pipe is not exactly right.

If a lead pipe is not exactly right for the flange to fit, it may be

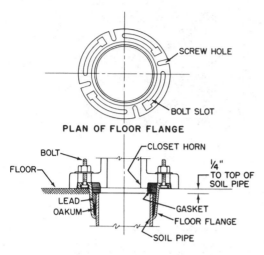

SCREW HOLE

BOLT SLOT

PLAN OF FLOOR FLANGE

BOLT
FLOOR
CLOSET HORN
1/4"
TO TOP OF
SOIL PIPE
LEAD
OAKUM
GASKET
FLOOR FLANGE
SOIL PIPE

FIGURE 5-10
To install a toilet, screw the flange to the
floor; then slip bolts into slots in the flange.

PUTTY
RING
PUTTY, RUBBER
OR WAX RING

FIGURE 5-11
With the flange in place,
fit a wax ring around the
horn of the toilet base
discharge opening. Then
apply either putty or plaster
(for ceramic floors) to the
edge of the bowl.

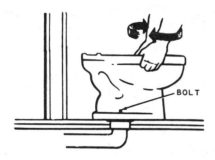

BOLT

FIGURE 5-12
Then the bowl is slipped over
the bolts and rotated slightly to
make certain it seats firmly.

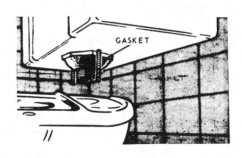

GASKET

FIGURE 5-13
The water closet is attached to
the bowl according to the
manufacturer's instructions.

dressed (bent) by tapping it gently with a dressing tool or with a hammer
handle to fit. Lead is soft, so fitting should be no problem.

In any case, the flange must be positioned so that you can locate
two bolts opposite each other. Then solder the flange (for watertightness)
to the pipe. For copper pipe, both flange and pipe first must be cleaned
with steel wool and coated with soldering paste. Then apply solder, a section at a time: Heat an area with a torch, melt the solder, and work your
way all the way around the pipe. Make certain that there is not a vein or
gap in the solder.

A lead pipe must also be cleaned before soldering. Tin (melt solder on, then wipe off) the edge of the flange with a torch and solder. Do not use a torch directly on the lead, because it will melt too easily. Final soldering may be done with a soldering iron.

Next, put the flange bolts in place and set up the bowl and tank dry to make certain that everything fits. Set the tank aside. Place a wax ring around the horn (the toilet discharge opening) and put the bowl in place, being careful to bring the bolts through the holes in the toilet. Rock the bowl gently to compress the wax, and work the bowl so it rests solidly on the floor. The bolts will protrude through the wax and the bolt holes. Tighten gently on both sides. (Do not use too much pressure, or you may crack the bowl.) Then put the elongated washers and nuts on the bolts. Install the tank according to manufacturer's directions, and hook up the water supply with a soft brass tank supply tubing and angle stop. A variety of possibilities are shown in Fig. 5-14.

LAVATORIES AND SINKS

There is much confusion in some people's minds as to the difference between a sink and a lavatory. Actually, a lavatory is just the technical name for a bathroom sink. When talking of a sink, plumbers are referring to a kitchen sink, service sink, or some other type.

Lavatories come in a wide variety of styles, colors, and sizes. Most lavatories are constructed of cast iron covered with porcelain, of vitreous china, or of steel covered with porcelain. Also obtaining acceptance is the plastic kind, which combines the lavatory and the counter top in a one-piece seamless unit.

The styles of lavatories most popular are the self-rimming, the under-the-counter, the wall-hung, and the flush mount models.

The flush mount kind needs a metal framing ring to hold it in place in the counter top. It is not costly, but cleaning between the frame and the counter can be a problem.

The self-rimming type has its own rim, which rests in a sealant applied to the counter top, with the rim projecting slightly above the countertop.

The wall-hung lavatory is just that—mounted to a special hanger secured on the wall.

The under-the-counter type is mounted beneath an opening cut in the counter top, with the fittings mounted through the counter top.

Kitchen sinks may be stainless steel, porcelain on steel, or porcelain on cast iron. Cast iron is better and more expensive than steel. Stainless steel comes in a wide range of quality and price. The most common

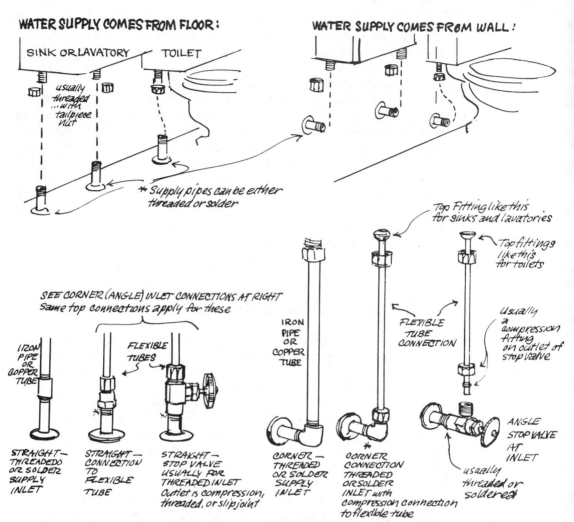

WATER SUPPLY COMES FROM FLOOR:

SINK OR LAVATORY   TOILET

usually threaded ...with tailpiece nut

*Supply pipes can be either threaded or solder

WATER SUPPLY COMES FROM WALL!

Top Fitting like this for sinks and lavatories

Top fittings like this for toilets

usually a compression fitting on outlet of stop valve

SEE CORNER (ANGLE) INLET CONNECTIONS AT RIGHT
Same top connections apply for these

IRON PIPE OR COPPER TUBE

FLEXIBLE TUBES

IRON PIPE OR COPPER TUBE

FLEXIBLE TUBE CONNECTION

ANGLE STOP VALVE AT INLET

usually threaded or soldered

STRAIGHT—THREADED OR SOLDER SUPPLY INLET

STRAIGHT—CONNECTION TO FLEXIBLE TUBE

STRAIGHT—STOP VALVE USUALLY FOR THREADED INLET Outlet is compression, threaded, or slipjoint

CORNER—THREADED OR SOLDER SUPPLY INLET

CORNER CONNECTION THREADED OR SOLDER INLET with compression connection to flexible tube

* Most of these fittings are available with either threaded or solder inlets

FIGURE 5-14

There are various ways to hook the toilet into the water supply. Techniques and materials used are similar to those used to hook a sink or lavatory into the water supply.

sink sizes are 21 × 24 inches in sinks that require a mounting rim and 22 × 25 inches in self-rimming sinks.

As with any sink (or tub or toilet, for that matter), fittings do not come with the unit and must be ordered and installed separately.

FIGURE 5-15
*Gaining new
popularity is the
bidet, which is of
European origin.*

installing a sink

If you have a choice, install the faucet on the sink before you put the sink in place. Sinks have three holes in which the faucet is mounted. The center hole is used for the spray hose or the pop-up drain fitting rod. There are other types of faucets that have the hot and cold water supplies that are both in the center hole and two screws to clamp the faucet in place in the other holes.

Some faucets come with plastic or rubber pads to seal them to the sink; others require putty around the holes.

If the faucet must be installed after the sink is in place, a basin wrench is necessary. This wrench will tighten or loosen a nut according to which way the jaws are turned. There are four positions for the jaws. Two give extra leverage. Two work well if you use one hand to close the jaw and the other to turn the wrench with the slide lever handle.

You can buy any type of faucet that goes into the drillings of the sink or basin. The most common drilling is 4-inch centers for a basin (lavatory) and 8-inch centers for a sink. Sinks sometimes have an extra hole for a spray. Both are commonly installed from the top with nuts underneath. However, there are others that work in the opposite way (with fancy handles held in place by escutcheons on top). This type may also be tightened from the bottom when in place.

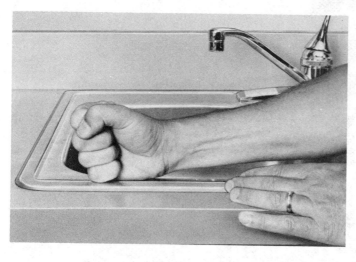

FIGURE 5-16

Some sinks can be installed with special clips that make for a secure friction fit when the sink is pounded into place.

FIGURE 5-17

A self-rimming lavatory.

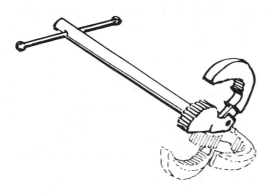

FIGURE 5-18

A basin wrench.

First, put the rubber gasket on the faucet wherever it comes into contact with the sink. Use plenty of putty if there is no gasket. Push it in place, put washers and nuts on from the underside; have a helper hold the faucet in position and tighten nuts with a basin wrench.

Next, attach the faucet supply pipe to the hot and cold water pipes. (These supply pipes are available in chrome or brass in lengths of 12, 18, 24, and 36 inches.) Slip the nuts over the supply pipe and screw to the faucet. Once the pipes are attached, bend them carefully so that they are parallel to the compression fittings. At least an inch of the pipe must be straight above the point of connection, or it will not seat properly. Mark the pipe and cut off excess so that about ⅜ inch will slip into the fitting. Slip the nuts and ferrules over the ends of the pipes, then slip the ends of the pipes into the fittings and carefully turn down on the nuts. It is important that the end of the pipe be aligned exactly in the fitting. Make certain that the nut is threaded properly before you tighten it.

Remove the aerator, if any, from the faucet. Turn on one valve at a time and check for leaks.

If the connections leak at the nuts, tighten them just enough to

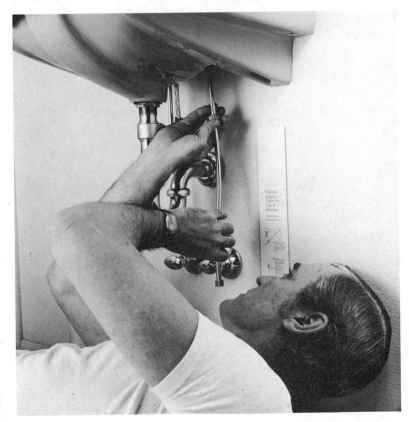

FIGURE 5-19

Supply pipes for the sink or lavatory are secured to the faucet shanks and supplies coming out of the wall. Cut these pipes to fit with a tubing cutter, making sure you don't crush them.

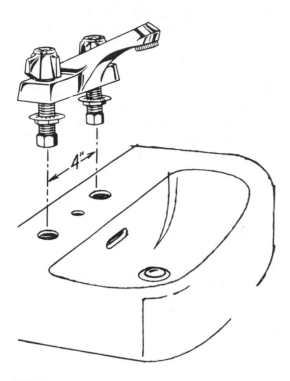

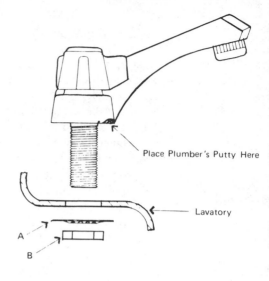

Place Plumber's Putty Here

Lavatory

A

B

1. Place plumber's putty in groove as indicated.

2. Insert shanks of Lavatory Faucet into holes of lavatory. Attach washer (A) and locknut (B) to shank. Tighten solidly in place, removing excess putty from base of faucet.

3. Connect water supply to shanks.

FIGURE 5-20
A 4" lavatory faucet (centerset) without pop-up drain. Measurement is made between centers.

FIGURE 5-21
Installing a faucet in the lavatory.

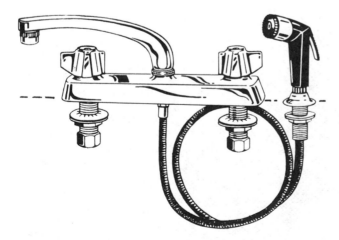

FIGURE 5-22
An 8" exposed deck faucet with spray. It has a 0-to-1¼" adjustment. This type of faucet is also available in 4" and 6" sizes.

stop the drip; use only open-end wrenches on compression fittings, and always use another wrench to "hold against yourself" so that you do not twist the pipe.

When the water supplies are tight, turn the water on full blast to check leakage at the waste connections. If they do not leak, put the plug in the sink and allow it to fill about 1½ inches deep. Then pull the plug and

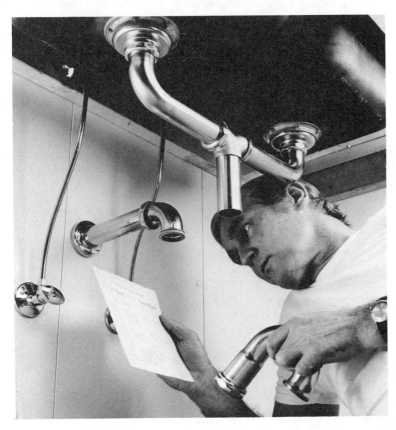

FIGURE 5-23

The trap is easy to install on sinks and lavatories, but you must make sure that the pipes are on straight. Otherwise, you'll get leaks.

FIGURE 5-24

There are special hangers for some types of sinks.

watch again for leaks. If there are any leaks, tighten nuts carefully. If it still leaks, disassemble the waste connection and do it over.

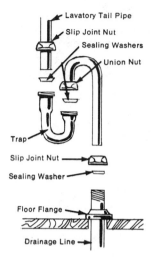

FIGURE 5-25

Some drain lines come up directly out of the floor instead of the wall.

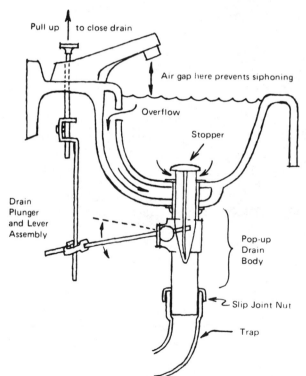

FIGURE 5-26

Operation of a typical lavatory with pop-up drain.

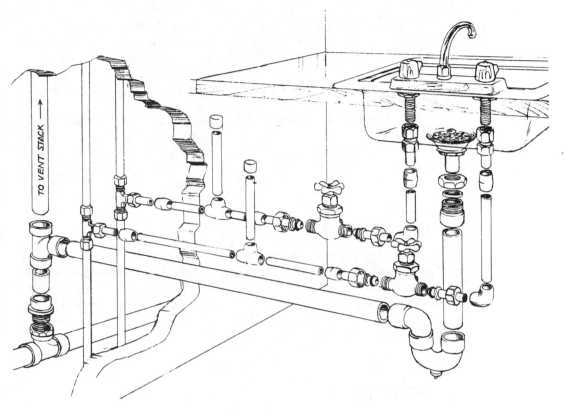

FIGURE 5-27
An overview of a sink and its connections.

installing a strainer and tail piece

A strainer and its corresponding tail piece are used on either a tub, basin, or sink. There are several types of strainer on the market. If you examine them before installation, you will soon see how they work.

The most common strainer is made up with a large rubber washer and a metal washer and nut under the fitting. Inside the fitting there is a chrome flange. This section is beaded in stainless putty.

To install it, take the inside section of the strainer in hand and press a bead of putty about ½ inch thick all around the underside of the flange. Press this section down into the hole in the fixture from the inside, and put the washers and nuts on from the outside. The excess putty will squeeze out. Have an assistant use the handles of a large pair of pliers inserted between the cross in the bottom of the strainer and a large screwdriver to hold the strainer from turning while you take up on the nut from below. Make certain that the rubber washer does not become displaced while you do this. If it does, release the pressure on the nut and replace the washer (it may be necessary to remove some putty from the underside of the fixture) and take up on the nut again just tightly enough that the strainer is snugly in place and does not easily turn from the pressure of the wrench.

The tail piece is connected to the strainer with a nut; there is a special washer that fits into the end of the tail piece.

The connection can also be made up with lamp wick as a slip joint, but the wicking must be put around the tail piece, not where the washer would have been.

installing a basin

A basin (lavatory) is installed just as a sink is, with the only difference being a pop-up drain. First, install the tail piece. Install the pop-up drain as you would a sink strainer, making sure that the bottom part is pointed in the right direction—straight back at the wall. The flange should be on the inside and beaded in putty. On most newer faucets the drain plug can be installed by twisting and pulling upward.

SHOWERS

See Figs. 5-28 through 5-31 regarding shower installation.

URINALS

The plumber is called upon to install a variety of urinals in public buildings. If one has a choice, the kind that should be recommended is one made of vitreous china or earthenware; wooden ones are absorbent and become fouled, and those made of iron get pitted by corrosive action.

Three kinds of urinals are common: trough, wall-hung and stall. The trough type may be 24, 36, 48, or 60 inches long, and is usually 8 inches deep and 12 inches wide, with the bottom of the trough approximately 22 inches from the floor.

Wall-hung urinals, like the trough type, have the bowl mounted on the wall. The shape, generally, is either round or lipped, with the lipped shape being preferable. Two methods of flushing are commonly used—washdown and siphon-jet. If the urinal is the lip type, the flushing rim is preferred, because its action cleans the bowl completely.

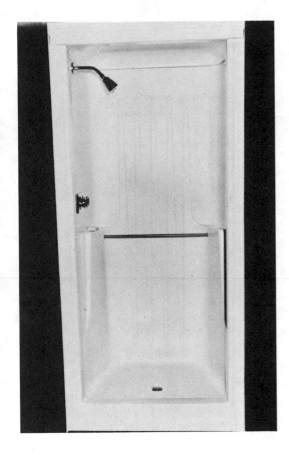

FIGURE 5-28
A one-piece wall shower.

The stall urinal is made of vitreous china, is set against the wall, and is sloped so that waste can drain easily. A rim-type flushing action cleanses the urinal after use. Usually, stall urinals come in braces of three, but they may be set together as needed, with the space sealed with cement, plaster of Paris, or some other waterproof, durable material.

DRINKING FOUNTAINS

Two types of fountains are most common: the wall and the pedestal types. As the name implies, the wall-hung fountain is mounted on a wall; the pedestal type is mounted on a floor. They are usually made of vitreous china in a wide variety of styles.

Installation of wall or pedestal drinking fountains varies greatly. Usually, the water supply is controlled by a valve at the spout opening and a self-closing compression stop. The valve avoids excessive waste of water.

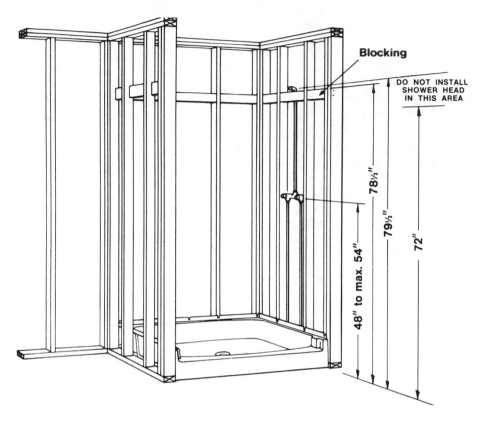

Blocking

DO NOT INSTALL
SHOWER HEAD
IN THIS AREA

78½"

79½"

72"

48" to max. 54"

FIGURE 5-29
Roughing in for a shower.

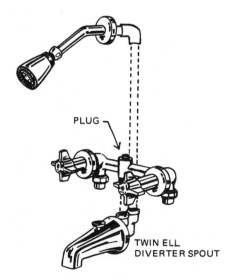

PLUG

TWIN ELL
DIVERTER SPOUT

FIGURE 5-30
*A two-valve diverter with shower head and twin ell
diverter spout.*

WOOD FLOOR

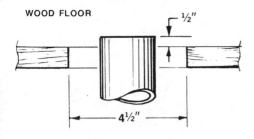

CONCRETE FLOOR

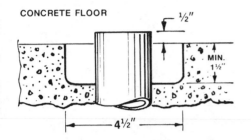

FIGURE 5-31
Installation of a shower drain.

Traps on these fixtures are usually no more than 1¼ inches. The fountain may be connected to the plumbing system proper, or not. For screw pipe drains a riser of 1½ inches is used; for soil pipe it should be 2 inches long.

An important part of the plumber's job is installing a drainage or DWV system. There are many variables in a task of this kind, because of the individual nature of a structure, the local plumbing codes, and, if the job is being done on an existing structure, what the plumbing set-up already is.

Following are general instructions for installing a typical drainage system in a home. Before we go into that, however, there are some general principles about the job that you should know.

GENERAL COMMENTS

For economy, the stack, which is the most expensive part of the job, should be a single unit and fixtures should be located as close to it as possible. The further away they are, the costlier the job. The stack extends from the building drain through the roof. Usually 3- or 4-inch pipe is used.

If there is one toilet, it must drain into the stack; extra toilets can drain into it if they are not too far away. The branch drain for the toilet must be the same size as the stack, and it must slope properly. If you need an additional stack for a toilet, it must be 3 to 4 inches in diameter (whatever is used in the main stack) where the toilet drains, and 2 inches in the vent portion. Alternatively, you can drain the extra toilet into the main stack if a 2-inch vent at the toilet is provided.

Other fixtures can drain into a main or secondary stack. Branch drains must be 1½- to 2-inch pipe, depending on the fixture, and, of course, they must slope properly.

6

Installing a Drainage System

Any secondary stack must also be vented properly (through the roof, or it may connect into the main stack at a point above the so-called flood run of the fixtures).

A branch drain that enters a stack 8 feet or more below another branch drain must be revented to prevent siphoning off water at the lower fixture trap. Revent pipe should be the same size as the branch drain.

All stacks must connect with the building drain, and this should be the same size as the main stack. The building drain must slope down all the way to disposal. There should be a trap and a fresh air inlet located where the soil pipe leaves the building.

TOILET DRAIN

The first step is to cut any necessary wall and floor openings for the stack and toilet lines. If the bathroom is above a room ceiling, cut away an 8-inch wide strip of flooring from the center line of where the toilet bowl outlet to the wall would be, then back through the wall and bottom partition header. This provides space for installing a stack tee fitting and closet bend assembly from above. If the ceiling is unfinished, the job is easier. You just cut one hole for the stack and one for the closet bend opening, and you can install the parts from below.

Next step is to assemble and install the toilet drain. Fittings for plastic pipe would be a tee, a closet bend, two short lengths of pipe, and a floor flange. If you use copper, use a ¼-inch bend fitting instead of a closet bend. To get the proper size for either assembly, first measure the length of pipe needed between the flange, which is secured to the finished floor, and the bend, so that the latter will be the proper depth.

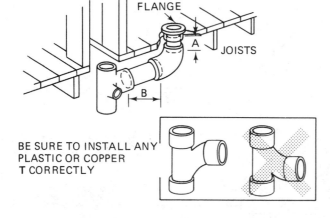

BE SURE TO INSTALL ANY PLASTIC OR COPPER T CORRECTLY

FIGURE 6-1

The toilet drain assembly for copper or plastic pipe consists of a T, a closet bend, and two short lengths of pipe. The lengths of pipe needed are indicated by A and B.

Next, join the bend and pipe length. Hold the tee and bend and measure the length of the horizontal pipe needed to the stack. Temporarily assemble the tee, horizontal pipe, and bend. To facilitate assembly you can lay the parts on the floor, braced so that the center lines are parallel, and mark them. Then join them as required, using the marks as guides.

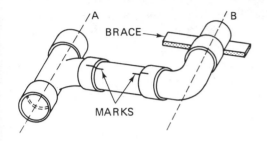

FIGURE 6-2

To join pipe, first temporarily assemble it on the floor, and mark where the pieces fit together; then permanently assemble it.

If you are working with cast-iron pipe fittings, the job is similar. If the assembly will go between joists, you need only a tee and a closet bend. Proper lengths can be achieved by breaking off parts of the toilet bend as needed. If you are going below joists, a tee, a hub-top long ¼-inch bend, and vertical pipe are needed; the other lengths needed are established as are those of plastic or copper. When all the pieces are ready, assemble them and mark them as before and brace them as needed to caulk the joints.

Next, nail the assembly into position temporarily. You should not install the floor flange until other work has been completed and you can finish the bathroom floor, because the flange will rest on the finished floor.

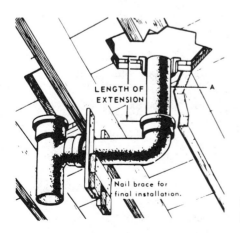

FIGURE 6-3

Cast iron pipe in position.

BUILDING DRAIN

Installation of this is next. If you have a plastic stack, either a 4-inch cast-iron or a 4-inch plastic pipe may be used for the underground building drain. If you have a copper or cast-iron stack, 4-inch cast iron is generally used. When you go from one type of pipe, say copper, to another, say cast iron, an adapter must be used. This is ordinarily located at the foot of the stack just above the ground or the basement floor.

Next, locate where the foot of the stack should be. Use a plumb bob and hang the string to the center of the toilet drain tee with the bob ½ inch from the floor. When the string is stationary, mark the floor where the bob would touch.

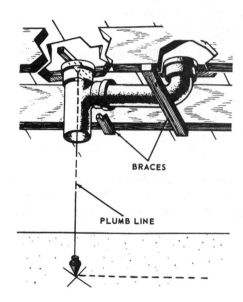

BRACES

PLUMB LINE

FIGURE 6-4

The foot of the stack can be located using a plumb bob as shown.

Next, from the mark draw a line for where the building drain will go out through the house foundation. Dig a 2-foot wide trench with this line as the center, deep enough so that the trench will be about one foot below the finished floor and graded from that point 1¼ inches every 5 feet. Make sure that the earth in the bottom of the trench is compact.

Then go outside and complete the trench, following the same grading and making certain that the earth is firmly packed, until the trench extends 5 feet beyond the foundation wall.

The building drain, as noted earlier, should have a cleanout assembly. Assemble this first. If it is to be plastic, the assembly can consist of a cleanout tee and two ⅛ bends, or a Y with a cleanout adapter and plug and a ⅛ bend. A cast iron cleanout assembly can be a test tee and two ⅛

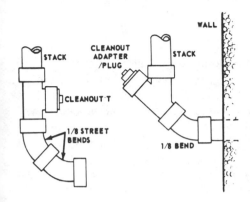

FIGURE 6-5

Fittings used for the cleanout assembly on a building drain.

bends or a Y with cleanout ferrule and plug and a ⅛ bend. In any case, align the parts in the same way you did for the toilet drain assembly and join as required.

Brace the assembly in the trench, centered under the plumb bob; then pour concrete all around and under it to set it in this position. Let the concrete set. Then install the rest of the building drain out to 5 feet beyond the house foundation. If branch drains were part of this, you would set in Y fittings at the same time. It is a good idea also to install all other under-floor drains now, so that the basement floor can be finished and cleanup accomplished before the installation is finished. A floor drain assembly is prepared and cradled in concrete in the same way as a cleanout. More about this later.

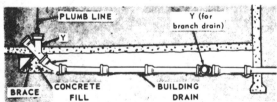

FIGURE 6-6

The building drain in final position, pitched and braced in concrete.

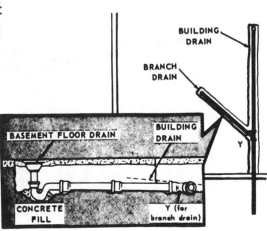

FIGURE 6-7

Y fittings and branch drains should be installed as the building drain is installed.

SUSPENDED DRAIN

In some instances an underground building drain will be impractical, and it will be necessary to install a hanging, or suspended, building drain. If the stack is plastic, this drain may be 3-inch plastic; with a copper stack, it may be 3-inch copper up to the Y assembly located in the basement wall or floor. With a cast-iron stack, 4-inch cast iron must be used. In most cases two cleanout assemblies using T-Y fittings are used, one at the foot of the stack and the other where the drain goes through the wall or floor; the drain from the second assembly outward is 4-inch plastic or 4-inch cast iron, the same as the underfloor drain. If only one cleanout assembly is used, all horizontal drains should be 4-inch pipe.

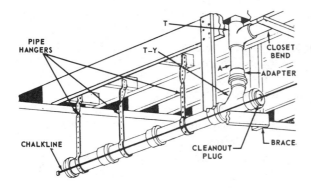

FIGURE 6-8
A suspended drain.

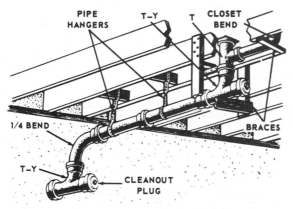

FIGURE 6-9
A building drain arrangement where there is a second cleanout in the wall.

Next, prepare the cleanout assembly that goes at the foot of the stack. If necessary, use a reducer or adapter at the top of the assembly for connection to the stack. Then join the assembly to the toilet drain assembly either directly or with a length of pipe between them. Length A may be needed if the drain must go through the wall at a certain height; it can be determined by using a chalk line as shown in Fig. 6-10. Brace this assembly firmly in place.

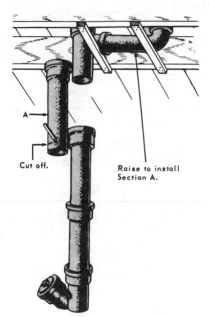

A→

Cut off.

Raise to install
Section A.

FIGURE 6-10

The last piece of stack can be held against the other stack parts, marked for cutting, and cut. You can raise the toilet drain assembly to fit it in.

Next, use a chalk line to determine where the suspended drain will go through the wall. It, too, should pitch at the rate of 1¼ inches per 5 feet. When you get to the wall, make a hole through it and prepare the trench outside for at least 5 feet out from the wall. Install the drain from the foot of the stack to this point outside the wall. If the drain includes a second cleanout assembly in the wall, preassemble the T-Y and ¼ bend and brace these into position to help you fit the pipe to them. Permanently brace each section of pipe every 5 feet to joists as you install it, remembering to keep proper pitch.

Next, install the rest of the stack from the cleanout assembly to the toilet assembly (this portion is already done if you have a suspended drain). The last piece should be held against the toilet assembly and stack, marked for cutting, and cut. The toilet drain assembly can be raised to fit the piece in.

Now, build the stack up from the toilet drain assembly and out through the roof. If there are any drainage or vent T's needed, install each as you come to it. Each branched drain tee should be at the proper height so that the drain will slope properly—1 inch every 4 feet. Revent runs should slope slightly upwards towards the stack. With a cast-iron stack, special vent T's are used. With plastic or copper, sanitary T's are used; these must be inverted in the stack when used for vent connections.

If there were a bathroom above the first one, you would first preassemble and temporarily install a toilet drain assembly as you did the first one, before building the stack up to it. This makes it easier to place the stack correctly.

In some instances obstructions will require turning the stack. This can be done by making an offset as shown in Fig. 6-11; use ⅛ vents if the offset occurs in the drainage portion of the stack and ¼ vent if it occurs in the vent portion. The stack should jut out about a foot above the roof and should be made absolutely watertight where it goes through. The flashing used around the stack is adjustable to roof pitch, and the roof shingles or other covering should overlap the flat part of the flashing at the top and sides. For watertightness seal the flashing to the stack pipe by peening the sides against the pipe.

FIGURE 6-11
If the stack must go around an obstruction —for example, a joist— you can use ¼ or ⅛ bend fittings.

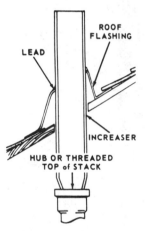

FIGURE 6-12
The method for sealing the stack where it exits through the roof.

Branch drains are installed out from the stack as needed. The drains may be plastic, copper, or steel and generally are 1½ inches in diameter; the same is true of any revent lines used. With plastic or copper, sanitary T's are used to join revent lines to the branch line, and the revent lines are built upward from these T's to join the inverted T's already installed in the stack. Steel pipe is used, either a vertical vent T or a horizontal vent T is used, and the vent run is built out and down from the stack T by caulking the unthreaded steel pipe end into the hub of the drainage line vent T. Two or more vent runs may be joined by a T or T's to run as a single line into the stack.

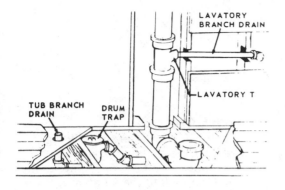

FIGURE 6-13

Branch drains should be installed on the stack as needed as you go.

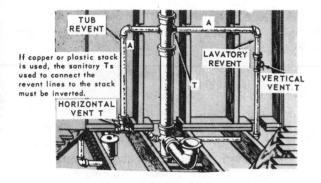

FIGURE 6-14

Installing sanitary Ts in copper or plastic stack.

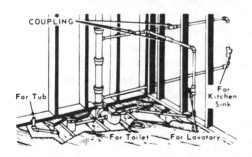

FIGURE 6-15

A vent run arrangement.

SECONDARY STACK

The secondary stack containing the toilet is installed just as the main stack is. The cleanout assembly must also be positioned and joined to the building drain as described. If there is no toilet, however, the cleanout location is not so critical. When installing the building drain and the branch drain to it from a secondary stack, simply hang a plumb bob approximately where the secondary stack will be and position the cleanout assembly accordingly. Build a secondary stack straight up from this clean-

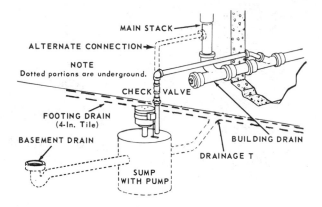

FIGURE 6-16

A setup for installing a building drain. If the building drain is suspended, the floor or footing drain must empty into the sump at a lower level.

out and install the branch drains, following the same general procedures described for a main stack.

INSTALLING A BASEMENT DRAIN

If final disposal is a city sewer, you can connect a basement floor drain or footing drain to the building drain. If the building drain is below the basement floor, you can make the connections through an underfloor branch. If the building drain is suspended, however, the floor or footing drain must empty into a sump at a lower level, and a sump pump must be installed to pump the drainage into the building drain.

The outside areaway drain or the roof drain can be connected with the building drain or building sewer if permitted. Otherwise, they should be drained off downhill or as described above.

If final disposal is a septic tank, you do not want to put any excess water into it, which can cause it to be overloaded. There are a variety of possibilities for installing a floor, footing, roof, or areaway drain. First, you can connect to the building sewer behind the septic tank. Second, you can drain the water into a separate disposal area—a stream or similar area, or a dry well. If you connect behind the septic tank, make sure that the disposal field is large enough to handle the volume of water. If a dry well is used, make it large enough to hold all the drainage from the biggest rainfall anticipated.

THE SEWER

The building sewer is that part of the piping of a building drainage system that goes from the building drain 5 feet outside the foundation wall to final disposal. Pipe used for it will depend on local building codes, but a number are in general use—cast iron, plastic, fiber, or vitrified tile. The minimum size is 4 inches. Of all the pipes, cast iron is the strongest and should be used wherever strength is required, such as under a driveway. However, if you can, use plastic or fiber. They are lighter and much easier to work with. Not many building codes accept them, however.

Generally, if the sewer passes within 50 feet of a well or suction line, or within 10 feet of a drinking water supply line under pressure, the sewer must be leakproof. For cast iron, this means that the joints must be leaded and caulked or hubless fittings used. If plastic or fiber pipe is used, joints must be welded and both pipes must be solid, not perforated. Also, a cleanout must be installed wherever the sewer pipe changes direction 45 degrees or more, or wherever the grade of the sewer changes 22½ degrees or more. If it is not possible to avoid such a bend, it should be made with two 45-degree ells or a long sweep ¼ bend. A cleanout is also a good idea within 5 feet of a septic tank if the tank is more than 20 feet away from the building. An inexpensive cleanout can be had by inserting a T in the line with its vertical leg extended to ground level and plugged with a brass cap. However, if the sewer is deeper than 4 feet, manhole construction is required for cleanout.

FIGURE 6-17

The correct way sewage should flow in the sewer pipe.

A minimum grade of 1 percent—1 foot fall per 100 feet, or ⅛ inch per foot—is required. A fall of ¼ inch per foot is desirable. Steeper grading is permissible, but the grade throughout the 10 feet immediately preceding the septic tank should not be more than ¼ inch per foot. Too steep a grade, contrary to what you might think, can cause clogging.

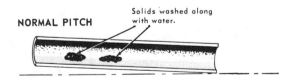

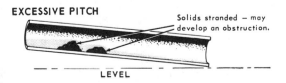

FIGURE 6-18

It is important to pitch sewer pipe correctly to ensure that waste moves along properly.

a trench for the sewer

The sewer pipe is, of course, underground, and a trench must be dug for it. The trench should be as straight as possible between the building drain and final disposal. It should be graded properly and deep enough to avoid frost and crushing, and wide enough that there is work space for making joints. To establish grade for the trench, first drive stakes into the ground 8 inches outside the trench width line and 10 feet apart. The stake by the foundation becomes a reference stake as shown in Fig. 6-19. Use a straight 10-foot board and level to level all other stakes driven. Set one end of the 10-foot board on stake 1, the other end against stake 2. Level the board and mark stake 2. Saw it off at the pencil line, thus making stakes 1 and 2 the same height.

Repeat the procedure, laying the board on stake 2, and holding it against stake 3. Now all three stakes are level. Continue until you reach the septic tank.

FIGURE 6-19
Grading and digging the trench for the sewer pipe and building drain.

grading the trench

To grade the bottom of the trench, you use a level and a 1 × 2 board. Set one end of the level on stake 1, holding the other end against the 1 × 2, as shown in Fig. 6-19. Make sure that the 1 × 2 is level with the bottom of the trench.

Mark the 1 × 2 with a pencil line (A). You are grading your sewer for ⅛ inch per foot, so you can measure up above the reference line 1¼ inches and make another line (B). This represents the depth of the trench at stake 2. Set the carpenter's level on stake 2; it should be level on the (B) line. Measure ¼ inch above the (B) line; this third (C) line represents the depth at stake 3.

If a grade of ¼ inch per foot is desired, the lines above the reference line (A) must be 2¼ inches apart. There should be a line marked on a 1 × 2 for each stake driven. It might be well to number each line corresponding to the stake numbers—1, 2, 3, 4, 5, etc.

To dig the trench more than one person can be involved, because the depth can be checked against the stakes. When laying the sewer, be sure that the ground is compacted. It is important that all digging be done to the proper depth from the start—no backfilling should be necessary.

using plastic pipe

As mentioned, there are advantages to using plastic pipe for a sewer line. Start at the building drain, using an adapter as required to join the first pipe length to the end of the drain. Add lengths as needed to extend the sewer to the septic tank. Make the final connection to the septic tank with a cast-iron pipe adapter, leaded and caulked into the tank inlet; the tank will have to be shifted when the final joint is made. All straight lengths of pipe are joined with couplings; bends can be made by using ¼-inch or, preferably, ⅛-inch bends, and either a T or Y fitting with a ⅛-inch bend may be used wherever there is a cleanout. The bottom of the trench must be grooved under each fitting to allow the pipe to lie flat on the trench bottom.

Finally, when all the joints have hardened, backfill the trench, compacting the soil thoroughly, up to ground level.

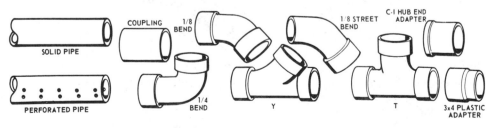

FIGURE 6-20
Plastic drainage pipe and pipe fittings.

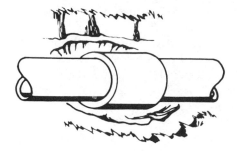

FIGURE 6-21
The trench must be grooved under pipe fittings to allow room for them.

using fiber pipe

Fiber pipe has all the advantages of plastic (light, easy to work), but the joints are not quite as easily made. (More on this later.)

As with plastic pipe, start at the house building drain, using a cut-off pipe and an adapter. Use 10-foot solid pipe lengths and couplings. The pipe ends are tapered, and so are the recesses in each coupling. Joints are made by driving coupling and pipe length together; the friction created heats the material enough to weld the joints. To drive, preassemble pieces so that the block of wood can be placed against the coupling (never against the pipe end) and struck with a sledge hammer. Each joint must be driven until the tapered pipe end is fully within but not beyond its coupling recess. One end of each ⅛ bend, ¼ bend, or Y is tapered outside like the pipe end to receive a coupling, and the other ends are shaped like a coupling to receive a tapered pipe end. The T fitting has all three ends recessed to receive pipe ends, so that all can be driven into place as if they were couplings. However, since they cannot be turned after assembly, they must be carefully aligned to point correctly while being driven into place.

FIGURE 6-22
Couplings for fiber pipe are pounded in place with a hammer and a block of wood.

FIGURE 6-23
A snap coupling for fiber pipe.

If a pipe is cut, the end must be tapered with a special tool for joining with a coupling, or the untapered end can be joined to a tapered pipe end by using an adapter. Also, two untapered pipe ends can be joined with a sleeve. However, any untapered fiber pipe joint must be sealed with special cement. It is also possible to join this pipe with cast iron; these joints must also be sealed. The final joint between the fiber and the septic tank inlet is made by similarly cementing an untapered pipe end into the inlet hub.

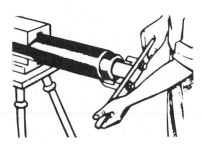

FIGURE 6-24
Tapering the end of fiber pipe with a special tool.

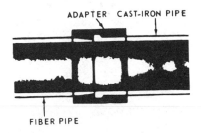

FIGURE 6-25
An adapter for joining fiber pipe to cast iron pipe.

using cast-iron pipe

Cast-iron pipe requires more room for joining; therefore, a wider trench must be dug. Every joint must be sealed, first with a ring of oakum and then with lead properly caulked in, or the hubless system using neoprene washers may be used. The direction of the flow must also be carefully observed to prevent clogging.

The oakum is yarned into the space between spigot and hub to a depth that will leave approximately 1 inch for the lead (actually, ¾ to 1 inch if hot lead is used; 1¼ inch if lead wool is used). It must be tightly and uniformly packed, so that the spigot is centered in the hub.

Hot lead is preferred, but lead wool may be used. Hot lead is melted in a plumber's furnace, with a gasoline blowtorch, or even over the kitchen range or a bonfire. It must be thoroughly molten for ladling into a joint, and the ladle must also be preheated. (A cold ladle dipped into hot lead can cause a minor explosion.) Also, a joint runner must be installed around the pipe to seal the hub end so that lead can be poured into the V opening at the top of the runner—into the hub—without running out. After the lead has set, remove the runner and caulk the lead firmly to squeeze it: (1) against spigot end; (2) against the inside of the hub. A caulking iron

FIGURE 6-26

Rigging cast iron pipe for pouring lead.

is used, and lead must be firmly packed all around against each of these surfaces. If lead wool is used, it is put in cold, by hand, in strands to the depth required to fill the hub. Each strand must be very thoroughly caulked in place until it has the appearance of solid lead.

disposal system

Sewage from the house normally empties into either a municipal sewer, a septic tank, or a cesspool. At its best, the cesspool causes a lot of problems. If there is a choice, a septic tank is the far better choice. The municipal sewer is, of course, better than either.

how a septic tank works

A septic tank receives raw sewage from the house. Inside, bacteria dissolve some waste, leaving a clear liquid on top (with some solids settling to the bottom of the tank), which flows from the tank through pipes that are perforated and is dispersed in the soil.

Again, as with most plumbing work, careful adherence to local plumbing codes is required in installing the tank and the field. All drainage lines in the field should be separated by a minimum spacing of 6 feet. Greater spacing is desirable if possible. All the so-called lateral pipes of the field should be an equal length to distribute the effluent (liquid sewage) equally. Under no conditions should the field have less than 150 square feet of effective absorption area. Pipes should not exceed a distance of 150 feet; also, at least two lines of pipe should be used. The design is intimately related to the amount of effluent that a particular soil can absorb.

The disposal field should also be graded for a number of reasons, including preventing standing water and soil saturation. The grade of a field laterally may vary from 0 to 6 inches per 100 feet but should never exceed 6 inches. A level grade is preferred. It is desirable to have lines 18 inches below the finished grade. In many cases soil conditions make it necessary to vary the depth of cover in order to maintain an even grade.

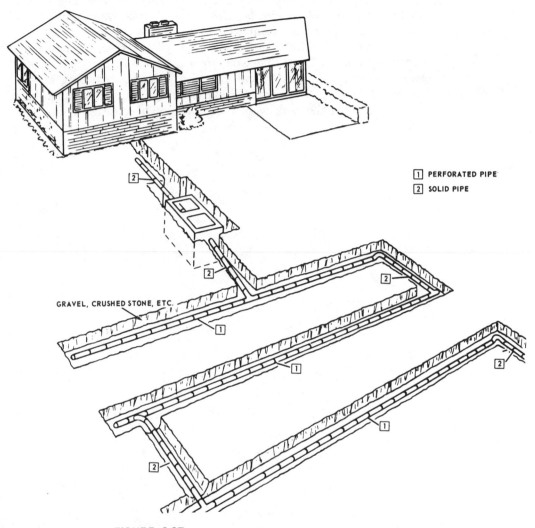

PERFORATED PIPE
SOLID PIPE

GRAVEL, CRUSHED STONE, ETC.

FIGURE 6-27
A good setup for a drainage field and sewer pipe.

However, the total depth of any lateral trench should not be more than 36 inches.

maintenance of a septic tank

Cleaning a septic tank depends on how many people are in a house, but ordinarily it should be pumped out only every 2 or 3 years. It is important, also, to keep foreign substances out of the tank, such as filters on cigarettes, which do not dissolve but clog the lines. Rain water should also be routed so that it does not saturate the field.

Every 18 months or so, the tank should be inspected and the depth of the accumulated sludge determined. This is done with a long pole. When the sludge reaches a depth of 18 to 20 inches, the tank should be cleaned. It is better to remove the sludge in the spring, rather than the fall, to avoid loading the tank with undigested solids during the cold winter months.

Rule number 1 for installing the water supply system is to install the service line in as straight a line as possible from the water source to the most convenient point of entry into the building. The line should be deep enough that traffic does not disturb it and it is below the frost line.

If there is a private water source —say, a well—the house service line is considered to be all the piping up to the pressure tank from which internal main supply lines lead to the various branch supply lines and fixtures. Pipe size should be large enough to provide all the water anticipated to be necessary at peak load times.

All the pipes to the horizontal should have a slight slope back to the house service line, and there should be a stop and waste valve at the low point, so that the system can be drained if necessary. Pipe sizes, as mentioned, should be of a sufficient size to serve all fixtures properly. Generally, ¾- or 1-inch pipe is used for the hot and cold water systems and ½-inch pipe for branch lines to fixtures.

Gate valves should be installed wherever convenient and on both lines at the water heater (hot coming in, cold going out). It is also a good idea to install union couplings wherever the line may have to be taken apart, such as the lines to the hot water heater.

Normally, main supply lines are installed under first-floor joists, secured by hangers. Pipes can run at right angles or parallel to joists. Union L's and T's can be used to advantage in connecting branch lines to mains, because they eliminate the need for careful measuring and fitting.

Both hot and cold water lines

7

Installing a Water Supply System

should run parallel, spaced at least 6 inches apart to prevent the cold water lines from absorbing heat from the hot water lines; you can forget this requirement if the hot water pipes are insulated. The hot water line should be run to the left side of each fixture as you face that fixture.

CONNECTIONS TO THE FLOOR

Run branch lines from the hot and cold water main lines below the floor joists to a first-floor bathroom. Fasten the pipes to joists with pipe straps. If the pipes cannot be run below the joists—for example if a bathroom is on the second floor—run them below the floor and across the tops of joists, notching each joist to recess the pipe. Use plugs and caps to stopper the branch line ends while attaching to walls and floors.

Figure 7-1 shows a typical installation, taking the pipes through the walls. The supply lines are run under the joists, and the partition plate is notched for the vertical branch line to the fixtures. If the bathroom is on the second floor, supply lines are generally run up next to the stack. The supply lines are run horizontally through notches in the wall studs.

As mentioned previously, using air chambers is a good idea. They will prevent the common problem of water hammer.

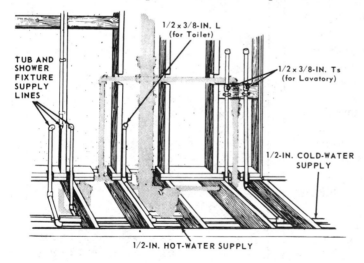

FIGURE 7-1
An X-ray view of a water supply system.

WELLS

In rural areas and vacation areas outside cities where municipal water is not available, the only good source of water is the well. It is well, one might say, to know something about them.

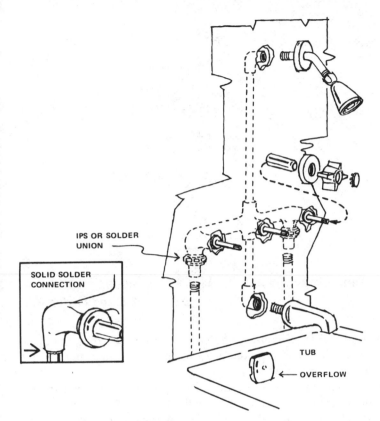

IPS OR SOLDER
UNION

SOLID SOLDER
CONNECTION

TUB

← OVERFLOW

FIGURE 7-2
*A water supply system linkup
with shower fittings.*

Who may drill a well and where it may be located should be
strictly limited by local codes. It is rather easy for water to become contaminated to the point where it can cause illness, even death. Unfortunately,
even water that is clean-looking and pleasant to the taste can be contaminated, and it should always be tested before being used as a water
supply.

different kinds of wells

There are a variety of well types. The simplest is the so-called driven well.
This is created by sledge hammering a pointed pipe into the ground. The
pointed end also has a screen. The pipe, usually in coupled sections, is
driven into the ground until the screen is below the water table level.
Water then enters the pipe, with the screen filtering out sediment, and a
pump draws it up.

Usually, a driven well is good for areas whose ground is composed
of relatively coarse sand. The whole operation can be accomplished very
cheaply.

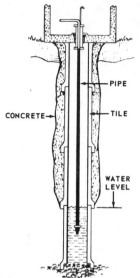

CONCRETE → | ← PIPE
| ← TILE
WATER LEVEL

FIGURE 7-3
A driven well.

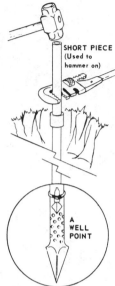

SHORT PIECE
(Used to hammer on)

A WELL POINT

FIGURE 7-4
A bored or dug well.

A driven well is, of course, limited by the terrain. If the point of the pipe hits rock, the well stops there. Consequently, you cannot install this type of well in many areas.

Another type of well is the dug well. These can be constructed with either hand or power tools and are usually 3 to 4 feet in diameter, but they may be much larger. Usually, they are less than fifty feet deep, but they must sufficiently penetrate the water table. They normally consist of concrete well rings, 3 feet in diameter and 2 feet deep. The well hole is made a foot larger in diameter than the well ring, and the upper part of the space is filled with concrete.

Bored wells are very much like dug wells, but they may be deeper and are usually smaller in diameter. They are dug with either hand or power augers.

Hand-bored wells are usually 8 inches or less in diameter; power-bored wells may be as large as three feet.

Bored wells are usually encased for their entire length. Some may require a screen for proper entrance of water.

Although a well is simple to construct, it does take skill and experience. Too often the hole collapses, and the well digger loses his equipment.

Drilled wells are the best wells. The equipment used can go to great depths through rock formations, and a supply of adequate water is almost guaranteed. For domestic use, the wells are normally six inches in diameter.

pumps

Most wells use pumps to bring the water to the surface, and there are a variety of these. Some are designed for drawing water from relatively shallow wells, others from medium to deep wells.

The piston-type pump is for shallow wells; both hand and automatic kinds are available. The simplest kind consists of a cylinder, a piston, and two valves. As the piston rises, a check valve at the bottom of the cylinder is opened, and, at first, air, then water is drawn into the cylinder. As the piston descends, the check valve is closed and the second valve, on top of the piston, opens to let water pass through the piston. Successive strokes result in more and more water being pumped until continuous flow occurs.

The automatic piston type has a double-acting piston that, by virtue of four valves, simultaneously draws and pushes water on both up and down strokes—a sturdier stream is delivered.

The centrifugal pump, also for shallow wells, has an impeller designed to move water rather than air. Water is sucked into the eye of the impeller, which picks it up and whirls it through the pump discharge opening. If all the well-pipe air is first disposed of, the flow of water becomes continuous, with force depending on the size of the impeller. A centrifugal pump must be primed to start, or kept primed by a foot valve.

medium to deep well pumps

A centrifugal jet-type pump is a centrifugal pump with a "jet" added. The jet serves three functions: It diverts some of the discharge water back into the impeller; a jet nozzle increases the pressure of the diverted water, and it draws more water from the well.

The submersible pump is designed for wells up to 500 feet deep, but it can also be used in wells as shallow as 20 feet. Its great advantage is that it is a compact unit with both pump and motor submerged in water in the well. There is no need to have a basement or a separate pump house installation. Only a tank and a small control box are required above ground, and they can be conveniently located in a utility room or corner of the kitchen. There is no pump noise or vibration.

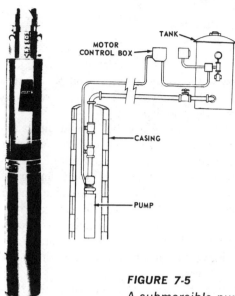

FIGURE 7-5

A submersible pump.

A plumber's job includes the maintenance and installation of water heaters. Basically, a water heater consists of a storage tank and a means of heating it. In operation, water comes in at the top of the tank through a dip tube and is routed to the bottom, where it mixes with already heated water. This process allows it to mix gradually rather than rapidly, which would bring down the temperature of the water too fast.

A thermostat on the tank keeps the hot water at a certain temperature. The water enters the hot water supply line through an opening at the top of the tank. There is also a magnesium anode rod in the tank whose purpose is to retard corrosion (by attracting it), thereby keeping other parts of the tank relatively corrosion-free.

At the bottom of the tank there is a spigot for draining off water and any scale or soil buildup. For safety, a hot water heater should be equipped with a temperature/pressure relief valve; if the water temperature gets too high, this will bleed off excess pressure. The newest water heaters also have an ECO (energy cut-off) device that shuts off the fuel supply—and therefore the fire—if water temperature gets above 200°F. The tank itself is insulated with fiber glass or some other material.

Water heaters are commonly fueled by natural gas, liquified petroleum gas, electricity, or oil. Let us take a look at each.

NATURAL GAS WATER HEATER   This type of heater has a flue area running through it or around it. A gas flame burns below the tank to heat the water, and the flue is vented to the outdoors.

8

Water Heaters

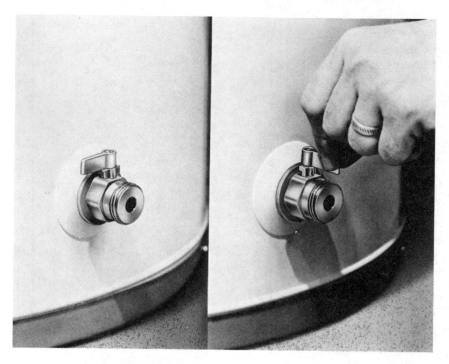

FIGURE 8-1

Water heaters like this A.O. Smith Deluxe 111 model have spigots at the base for draining off water. This one requires only a quarter turn.

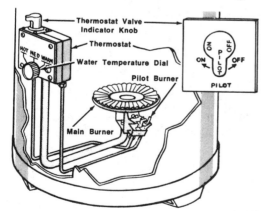

FIGURE 8-2

An X-ray view of the burner inside the water heater.

When the thermostat calls for heat, the pilot flame ignites the burner and it runs until the water reaches the proper temperature. A safety feature of this type of heater is that if the pilot flame goes out, all gas shuts down. Without this feature (some older heaters do not have it), the gas could escape through the house.

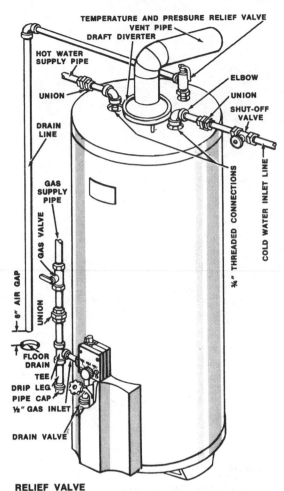

TEMPERATURE AND PRESSURE RELIEF VALVE
VENT PIPE
DRAFT DIVERTER

HOT WATER
SUPPLY PIPE

ELBOW

UNION

UNION

SHUT-OFF
VALVE

DRAIN
LINE

GAS
SUPPLY
PIPE

GAS VALVE

COLD WATER INLET LINE

¾" THREADED CONNECTIONS

8" AIR GAP

UNION

FLOOR
DRAIN

TEE

DRIP LEG

PIPE CAP

½" GAS INLET

DRAIN VALVE

RELIEF VALVE
NOTE: A new relief valve must be
installed. Be sure location complies with local codes. Shown is
typical relief valve location.

FIGURE 8-3
A gas water heater.

LIQUEFIED PETROLEUM GAS HEATER   This heater operates essentially
as does the natural gas heater, except that fuel and heating orifices are
smaller because the gas is much more concentrated. It is stored as a liquid
but vaporizes to burn. This type of heater has long been equipped with a
100 percent shutoff valve. Any gas heater should have a "Blue Star" cer-
tificate that certifies that the manufacturer has made it according to recognized safety standards.

ELECTRIC HOT WATER HEATER   Here, the tank is heated by an insulated
heating element immersed in the water. It, too, has a thermostat that calls

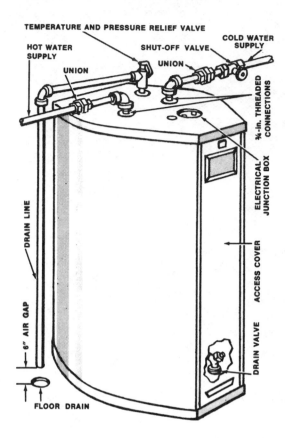

TEMPERATURE AND PRESSURE RELIEF VALVE

HOT WATER SUPPLY

SHUT-OFF VALVE

COLD WATER SUPPLY

UNION

UNION

¾-In. THREADED CONNECTIONS

ELECTRICAL JUNCTION BOX

DRAIN LINE

ACCESS COVER

6" AIR GAP

DRAIN VALVE

FLOOR DRAIN

RELIEF VALVE
NOTE: A new relief valve must be installed. Be sure location complies with local codes. Shown is typical relief valve location.

FIGURE 8-4
An electric water heater.

for heat and shuts off when the water is the proper temperature. Electric heaters may also have two elements controlled by individual thermostats.

For adequate hot water, an electric heater should be wired to a 240-volt electrical service. Some electric heaters operate on a 120-volt service, but they do not give much hot water. An electric heater does not require a flue, and it is normally cheaper than other types. However, since electrical costs have risen markedly in recent years, careful consideration should be given before purchasing one.

OIL HOT WATER HEATER   This type of heater uses a pressure-fire burner like the one on a furnace. The older types use a pot-type burner. There are rather complicated controls to start and stop the fire safely.

QUALITY

There are a number of things to look for when one is shopping for a water heater. Usually the best kinds are glass-lined. This keeps the water from eating into the sides of the tank. However, its usefulness really depends on the kind of water used, and it is something that should be checked out with a dealer.

Another criterion of quality is the willingness of the manufacturer to stand behind the heater. Top-quality ones are guaranteed for 10 years.

Electric and gas heaters come in a variety of shapes, including ones that can fit under a counter where space is at a premium.

SIZE OF THE HEATER

Residential water heaters are available in sizes from 10 to 80 gallon capacities. However, mere size is not the crucial factor. What engineers call the *recovery rate* is far more important. A nontechnical definition of this term is the ability of the heater to deliver water with the tap left wide open. Another way of putting it is to say that the recovery rate is how many gallons of water a heater will raise 100 degrees F in temperature in an hour. For example, if cold water enters the heater at 50°F and is heated to 150°F at the rate of 60 gallons per hour, the heater's recovery rate is 60 gallons per hour.

Make certain when checking the recovery rate that you are not comparing apples with oranges. Some manufacturers list a 60-degree rather than a 100-degree recovery rate.

Now, the actual amount of water that can be drawn from a heater in one hour equals the recovery rate plus about 70 percent of its storage capacity. When more than 70 percent of the stored water is drawn, cold water mixing with it lowers the temperature.

INSTALLATION

There are a variety of ways to install a water heater; the method of installation depends to a large degree on local codes. If possible, both the hot and cold water lines and the gas piping, if any, should be put in with corrugated brass connectors. They come from a foot to a foot and a half long with appropriate connectors and are very easy to handle. Gas should be transmitted to the water heater by black or galvanized steel; no ffexible connectors should be used.

If the heater is not electric, it should be located as near as possible to the flue—no more than 15 feet away. The flue connections should be at least as large as the heater flue outlet. A gas heater may share a good-sized chimney with a gas heating plant, but the heater flue should go into the chimney above the furnace flue. Any horizontal pipes should slope upward ¼ inch per foot.

Gas heaters should have draft diverters in the flue, and oil heaters need draft regulators. An oil heater works better when it has a separate opening into the chimney for the flue. Also all right is a Y-connection into the heating plant flue, but in this case each flue needs a separate draft regulator. A T-connection should never be used.

If the code permits, every water heater should have a temperature and pressure (T and P) relief valve and this should exceed or equal the heater's BTU output. (BTU = the amount of heat necessary to raise one gallon of water 1°C.) The element of the T and P valve should be in contact with the water at the top 6 inches of the tank. Most new heaters have a tapped opening for the valve. If yours does not, a tee may be installed next to the tank. The T and P valve should also have a discharge tube leading down from it to the basement floor or outdoors.

MAINTENANCE

The first rule for proper maintenance of a water heater is to set the temperature of the water properly. The manufacturer will recommend a figure, but it is normally 140°F. Too high a setting will result in your paying for hot water that you do not need and will shorten the life of the tank.

Once a month the heater should be drained by turning the petcock valve at the bottom of the tank. Draining draws off sediment in the tank and prevents a buildup inside, which can interfere with the heater's functioning. It is also a good idea to lift the lever on the T and P valve to bleed off some water to check for safe operation.

If a heater is more than three years old, the magnesium anode rod in it should be checked. If it has seen better days (i.e., if it is covered with scale) get a new one. The same goes for an electric heater. Its element should be periodically unscrewed and visually checked for scale. Scaling usually occurs in hard water areas.

An oil heater should be cleaned and adjusted when dirty or every few years. When a fired heater is checked, the flue should be examined for efficiency. This examination can be made by holding a lighted match in the draft diverter or draft regulator and seeing if the flame is drawn by suction.

PROBLEMS

Though a water heater is a relatively stable appliance, some problems do develop. Following are symptoms of problems and possible solutions.

1. If the heater makes a cooking sound, it may be that scale has drifted down and built up a crust on the bottom of the tank. Droplets of water get under this crust and make the sound when they flash into steam. It is nothing to be concerned about.

2. If a faucet is turned on and there is a long wait for the water, the problem usually is in the heating system and can be cured by a circulating hot water system. Originally expensive, it can be worth the expenditure in the long run.

3. If there never seems to be enough hot water, it could be that the heater is too small, the thermostat is at too low a setting, the thermostat is defective, or the pipes are too small or improperly installed. The simplest solution here is to set the thermostat up higher and see how it works.

4. If water is rusty, it usually means that mud or silt has gotten into the tank. (The tank is not rusting.) It quite commonly occurs when a city periodically turns on the hydrants in a particular area. Turn the heat off and try draining the tank.

5. If there is a sizzling noise when the heater is on, it could mean that the tank is leaking or a nearby pipe is. If the tank is leaking you can try to repair it, as indicated in Chapter 10, or shop for a new heater.

6. If the T and P relief valve keeps popping, there may be a valve somewhere in the plumbing that is defective and is allowing excess pressure to build up when the heated water is expanded. To correct this problem, you can install an expansion chamber. Another possible cause is too high a temperature setting or a bad relief valve. For a temporary cure, lower the thermostat.

The plumber gets involved with the installation of a variety of heating systems that utilize water as a heating source. Following is a roundup of the most common systems, how they work, and tips on making them work better.

HOT-WATER AND STEAM HEATING

Hot-water and steam heating systems consist of a boiler, pipes, and room heating units (radiators or convectors). Hot water or steam, heated or generated in the boiler, is circulated through the pipes to the radiators or convectors, where the heat is transferred to the room air.

Boilers are made of cast iron or steel and are designed for burning coal, gas, or oil. Cast-iron boilers are more resistant to corrosion than steel ones. Corrosive water can be improved with chemicals. Proper water treatment can greatly prolong the life of steel boiler tubes.

Buy only a certified boiler. Certified cast-iron boilers are stamped "I-B-R" (Institute of Boiler and Radiator Manufacturers); steel boilers are stamped "SBI" (Steel Boiler Institute). Most boilers are rated (on the nameplate) for both hot water and steam.

Conventional radiators are set on the floor or mounted on the wall. The newer types may be recessed in the wall. Insulation should be installed behind recessed radiators, with either insulation board, a sheet of reflective insulation, or both.

Radiators may be partially or fully enclosed in a cabinet. A full cabi-

9

Heating
Systems

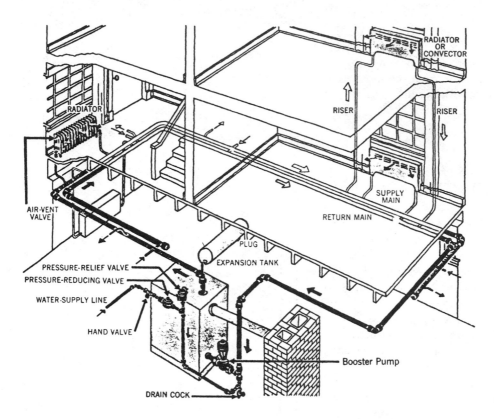

FIGURE 9-1

*Two-pipe forced-hot-water systems have two supply pipes or mains.
One supplies hot water to the room heating units; the other returns
cooled water to the boiler.*

net must have openings at top and bottom for air circulation. Preferred
location for radiators is under a window.

Baseboard radiators are hollow or finned units that resemble and
replace the conventional wooden baseboard along outside walls. They heat
a well-insulated room uniformly, with little temperature difference between
floor and ceiling.

Convectors usually consist of finned tubes enclosed in a cabinet
with openings at the top and bottom. Hot water or steam circulates
through the tubes. Air comes in at the bottom of the cabinet, is heated
by the tubes, and goes out the top. Some units have fans for forced air
circulation. With this type of convector, summer cooling may be provided
by adding a chiller and the necessary controls to the system. Convectors
are installed against an outside wall or recessed in the wall.

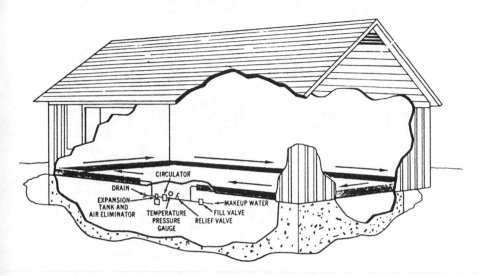

FIGURE 9-2

Electrically heated hydronic baseboard systems are made in units so that several units can be connected to form a single loop installation. Water is circulated by pump through the entire loop. Each unit has a separate heating element so that circulating water can be maintained at a uniform temperature, if desired, around the entire house.

FORCED-HOT-WATER HEATING SYSTEMS

Forced-hot-water heating systems are more efficient than the gravity hot-water heating systems.

In a forced-hot-water system, a small booster or circulating pump forces or circulates the hot water through the pipes to the room radiators or convectors.

In a one-pipe system, one pipe or main serves for both supply and return. It makes a complete circuit from the boiler and back again. Two risers extend from the main to each room heating unit. A two-pipe system has two pipes or mains. One carries the heated water to the room heating units; the other returns the cooled water to the boiler.

A one-pipe system, as the name indicates, takes less pipe than a two-pipe system. However, in the one-pipe system, cooled water from each radiator mixes with the hot water flowing through the main, and each succeeding radiator receives cooler water. Allowance must be made for this in sizing the radiators—larger ones may be required further along the system for proper heating.

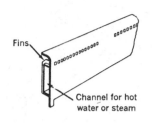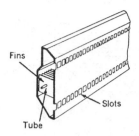

FIGURE 9-3

Baseboard radiator units are designed to replace conventional wood baseboards. In the hollow types (left and center) water or steam flows directly behind the baseboard face, and heat from the surface is transmitted to the room. In the finned tube type (right) water or steam flows through the tube and heats the tube and fins. Air passing over the fins is heated and delivered to the room through the slots.

Because water expands when heated, an expansion tank must be provided in the system. In an "open system," the tank is located above the highest point in the system and has an overflow pipe extending through the roof. In a "closed system," the tank is placed anywhere in the system, usually near the boiler. Half of the tank is filled with air, which compresses when the heated water expands. Higher water pressure can be used in a closed system than in an open one. Higher pressure raises the boiling point of the water, so higher temperatures can, therefore, be maintained without steam in the radiators, and smaller radiators can be used. There is almost no difference in fuel requirements.

With heating coils installed in the boiler or in a water heater connected to the boiler, a forced-hot-water system can be used to heat domestic water all year.

One boiler can supply hot water for several circulation heating systems. The house can be "zoned" so that temperatures of individual rooms or areas can be controlled independently. Remote areas, such as a garage, a workshop, or a small greenhouse, can be supplied with controlled heat.

Gas- and oil-fired boilers for hot-water heating are compact and are designed for installation in a closet, utility room, or similar space—on the first floor if desired.

Electrically heated hydronic (water) systems are especially compact, and the heat exchanger, expansion tank, and controls may be mounted on a wall. Some systems have thermostatically controlled electric heating components in the hydronic baseboard units, a type of installation which eliminates the central heating unit. Such a system may be a single-loop installation for circulating water by a pump, or it may be composed of individual sealed units filled with antifreeze solution. The sealed units depend on gravity flow of the solution in the unit. Each unit may have a thermostat, or several units may be controlled from a wall thermostat. An advantage of these types of systems is that heating capacity can be increased easily if the house is enlarged.

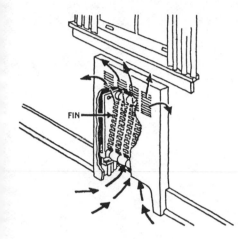

FIN

FIGURE 9-4

Convectors for hot water or steam are installed against the wall or recessed as shown.

STEAM CENTRAL-HEATING SYSTEMS

Steam heating systems are not used as much as forced-hot-water or warm-air systems. For one thing, they are less responsive to rapid changes in heat demand.

One-pipe steam heating systems cost about as much to install as one-pipe hot-water systems. Two-pipe systems are more expensive.

The heating plant must be below the lowest room heating unit, unless a pump is used to return the condensate to the boiler.

The plumber is called on to make a variety of repairs. Following is a roundup of the different types—common and uncommon—that occur.

FAUCETS AND VALVES

Faucets and globe valves, the type of shutoff valves commonly used in home water systems, are similar in construction and repair instructions given apply to both. Faucets or valves may differ somewhat in general design from the one shown in Fig. 10-1, because both come in a wide variety of styles.

Mixing faucets, which are found on sinks, laundry trays, and bathtubs, are actually two separate units with a common spout. Each unit is independently repaired.

If a faucet drips when closed or vibrates ("sings" or "flutters") when opened, the trouble is usually the washer at the lower end of the spindle. Water, which is constantly flowing up under pressure, is not completely sealed off. A little gets by and causes the drip. If it leaks around the spindle when opened, new packing is needed.

replacing the washer

1. Shut off the water at the shutoff valve nearest the particular faucet.

2. Disassemble the faucet—the handle, packing nut, packing, and spindle, in that order. You may have to set the handle back on the spindle and use it to unscrew and remove the spindle.

10

Repairs

147

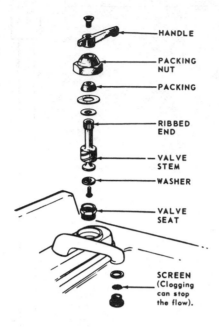

HANDLE

PACKING NUT

PACKING

RIBBED END

VALVE STEM

WASHER

VALVE SEAT

SCREEN (Clogging can stop the flow).

FIGURE 10-1

Parts of a typical faucet.

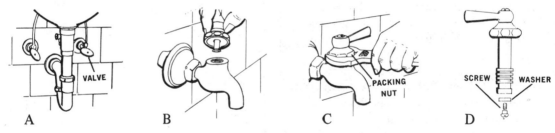

VALVE

PACKING NUT

SCREW   WASHER

A   B   C   D

FIGURE 10-2

A typical way to take a faucet apart: (A) First, turn off the water. (B) Loosen the packing nut; you can protect the chrome finish by first wrapping it in tape. (C) Turn the handle and the stem will lift out. (D) The washer is at the bottom of the stem, held in place by a screw.

3. Remove the screw and worn washer from the spindle. Scrape all the worn washer parts from the cup and install a new washer of the proper size.

4. Examine the seat on the faucet body. If it is nicked or rough, reface it with a seat dressing tool. Hold the tool vertically when refacing the seat; instructions come with the unit. If refacing fails, you may be able to unscrew the faucet seat with an Allen wrench and replace it with one of the same size.

5. Reassemble the faucet. Handles of mixing faucets should be in matched positions.

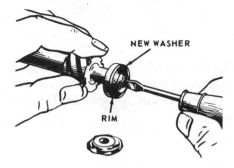

FIGURE 10-3

When removing the washer, use a screwdriver that fits the screw perfectly. It's a good idea to replace the screw with a new brass one. If the screw balks, you can chop away the washer and turn out the screw with pliers. Soaking the screw in penetrating oil facilitates removal; penetrating oil is good for loosening any parts.

FIGURE 10-4

If the rim around the bottom of the faucet stem is in good shape, you can use a type A washer (left) or a type B washer (center). If the stem is worn away, use a type C washer (right).

FIGURE 10-5

To remove some sink faucets, you must first take off the handle and then use the handle to turn and remove the stem.

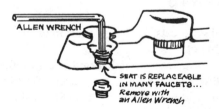

FIGURE 10-6

If replacing the washer doesn't solve the problem and the stem is in good shape, the seat may be at fault. This can be removed and replaced.

FIGURE 10-7

If the faucet chatters when you turn it on, the problem could be a loose washer or worn stem threads. If you replace the stem and it has play in it, this usually means that the threads are worn. A new stem is the solution. Occasionally a faucet will chatter and whistle because it is poorly designed. In this case a new faucet is the answer.

replacing the packing

To replace the packing, remove the handle, packing nut, and old packing, and install a new packing washer. If a packing washer is not available, you can wrap stranded graphite-asbestos wicking around the spindle. Turn the packing nut down tight against the wicking.

"O" RING

FIGURE 10-8
Modern faucets often use an "O" ring instead of packing.

replacing valves

After extended use and several repairs, some valves will no longer give tight shutoff and must be replaced. When replacement becomes necessary, it may be advisable to upgrade the quality with equipment having better flow characteristics and longer-life design and materials. In some cases, ball valves will deliver more water than globe valves. Some globe valves deliver more flow than others for identical pipe sizes. Y-pattern globe valves, in straight runs of pipe, have better flow characteristics than straight stop valves.

FROSTPROOF HYDRANTS

Frostproof hydrants are basically faucets, although they may differ somewhat in design from ordinary faucets.

Two important features of a frostproof hydrant are (1) the valve is installed under ground—below the frostline—to prevent freezing; (2) the valve is designed to drain the water from the hydrant when the valve is closed.

Figure 10-12 shows one type of frostproof hydrant. It works as follows: When the handle is raised, the piston rises, opening the valve. Water flows from the supply pipe into the cylinder, up through the riser, and out the spout. When the handle is pushed down, the piston goes down, closing the valve and stopping the flow of water. Water left in the hydrant flows out the drain tube into a small gravel-filled dry well or drain pit.

As with ordinary faucets, leakage will probably be the most common trouble encountered. Worn packing, gaskets, and washers can cause

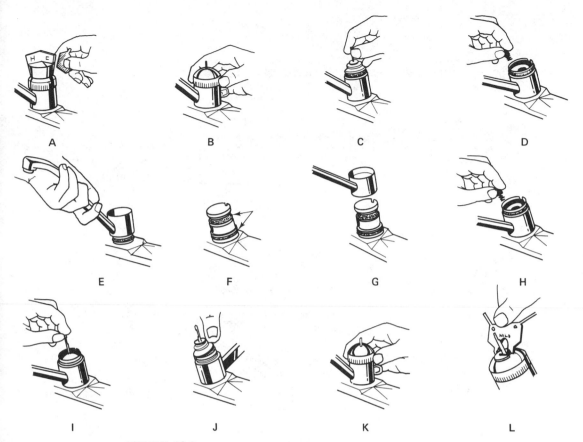

FIGURE 10-9

To service a single handle faucet: (A) Loosen the setscrew and lift off the handle. (B) Unscrew the bonnet assembly and lift it off. (C) Remove the retainer assembly and control cap by pushing sideways on the stem and lifting out. (D) Lift both seats and springs out of the sockets in the body. (E) Lift the spout straight up to remove it from the body. (F) Cut the spout seal rings at the points indicated and remove them from the body. (G) Stretch the seal rings and snap them into the grooves. Push the spout straight down over the body until it rests on the plastic slip ring. (H) Place the seat over the spring and insert it into the socket in the body (spring first). (I) Place the control cap into the body over the seats with the large flat of the stem to the rear on deck and lavatory models and down on shower models. (J) Insert the retainer assembly (with seal ring and packing) by first fully engaging the retainer tab in the body slot and then pushing down on the opposite side until the seal ring is fully inserted into the body. (K) Partially unscrew the brass adjusting ring. Then place the bonnet assembly over the stem and screw down hand-tight onto the body. (L) Screw down the adjusting ring until a light drag is felt when the stem is moved back and forth. Replace the handle and tighten the setscrew tight. Turn on the water supply.

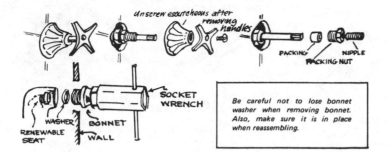

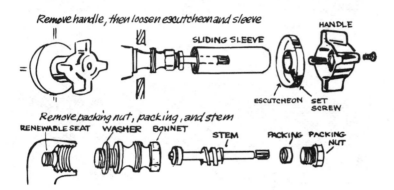

Be careful not to lose bonnet washer when removing bonnet. Also, make sure it is in place when reassembling.

FIGURE 10-10

Two types of bath and shower faucets.

FIGURE 10-11

The typical method of removing a shower faucet.

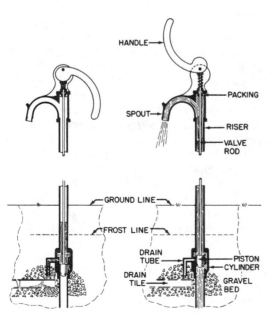

FIGURE 10-12

A frostproof hydrant: (A) closed; (B) opened. As soon as the hydrant is closed, water left in the riser drains out the drain tube as shown in A. This prevents water from freezing in the hydrant in cold weather.

leakage. Disassemble the hydrant as necessary to replace or repair these and other parts.

Frostproof yard hydrants having buried drains can be health hazards. The vacuum created by water flowing from the hydrant may draw in contaminated water standing above the hydrant drain level. Such hydrants should be used only where positive drainage can be provided.

Frostproof wall hydrants are the preferred type. For servicing sprayers using hazardous chemicals, hydrants having backflow protection should be used.

REPAIRING LEAKS IN PIPES AND TANKS

Leaks in pipes usually result from corrosion or from damage to the pipe; pipes may be damaged by freezing, by vibration caused by machinery operating nearby, by water hammer, or by animals bumping into the pipe. (Water hammer is discussed later.)

corrosion

Occasionally waters are encountered that corrode metal pipe and tubing. (Some acid soils also corrode metal pipe and tubing.)

The corrosion usually occurs, in varying degrees, along the entire length of pipe rather than at some particular point. An exception would be where dissimilar metals, such as copper and steel, are joined.

Treatment of the water may solve the problem of corrosion. Otherwise, you may have to replace the pipe with a type made of material that will be less subject to the corrosive action of the water.

It is good practice to get a chemical analysis of the water before selecting materials for a plumbing system.

repairing leaks in pipes

Pipes that are split by hard freezing must be replaced.

A leak at a threaded connection can often be stopped by unscrewing the fitting and applying a pipe joint compound that will seal the joint when the fitting is screwed back together.

Small leaks in a pipe can often be repaired with a rubber patch and metal clamp or sleeve. This must be considered as an emergency repair job and should be followed by permanent repair as soon as practicable.

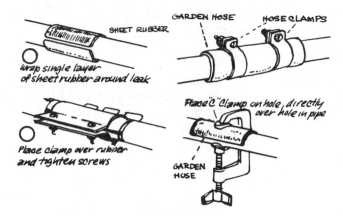

FIGURE 10-13

One temporary method of
stopping a leak in a pipe.

FIGURE 10-14

Other methods for stopping a leak. These
should also be considered temporary.

Large leaks in a pipe may require cutting out the damaged section
and installing a new piece of pipe. At least one union will be required, unless the leak is near the end of the pipe. You can make a temporary repair
with plastic or rubber tubing. The tubing must be strong enough to withstand the normal water pressure in the pipe. It should be slipped over the
open ends of the piping and fastened with pipe clamps or several turns of
wire.

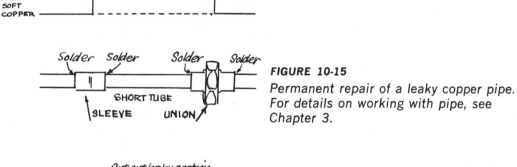

FIGURE 10-15

Permanent repair of a leaky copper pipe.
For details on working with pipe, see
Chapter 3.

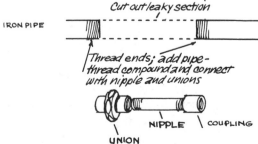

FIGURE 10-16

A method for permanently repairing iron
pipe.

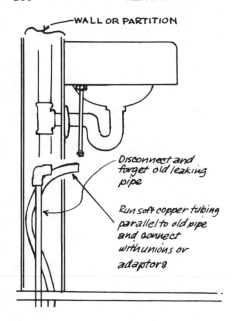

WALL OR PARTITION

Disconnect and
forget old leaking
pipe

Run soft copper tubing
parallel to old pipe
and connect
with unions or
adaptors

FIGURE 10-17

If the leaking pipe is behind a wall, the easiest method for repair is to plug up the old pipe and connect a new piece.

Vibration sometimes breaks solder joints in copper tubing, causing leaks. If the joint is accessible, clean and resolder it. (For tips on doing this, see Chapter 3.) The tubing must be dry before it can be heated to soldering temperature.

repairing leaks in tanks

Leaks in tanks are usually caused by corrosion. Sometimes, a safety valve may fail to open and the pressure developed will spring a leak.

Although a leak may occur at only one place in the tank wall, the wall may also be corroded thin in other places. Therefore, any repair should be considered as temporary, and the tank should be replaced as soon as possible.

A leak can be temporarily repaired with a toggle bolt, rubber gasket, and brass washer, as shown in Fig. 10-18. You may have to drill or

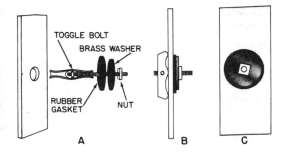

TOGGLE BOLT

BRASS WASHER

RUBBER GASKET   NUT

A   B   C

FIGURE 10-18

To plug a hole in a tank: (A) Pass the link of a toggle bolt through the hole in the tank, enlarging the hole if necessary. (B) Draw the nut up to compress the washer and gasket against the tank. (C) The completed repair.

ream the hole larger to insert the toggle bolt. Draw the bolt up tight to compress the rubber gasket against the tank wall.

WATER HAMMER

Water hammer sometimes occurs when a faucet is suddenly closed. When the flow of water is suddenly stopped, its kinetic energy is expended against the walls of the piping. This causes the piping to vibrate, and leaks or other damage may result.

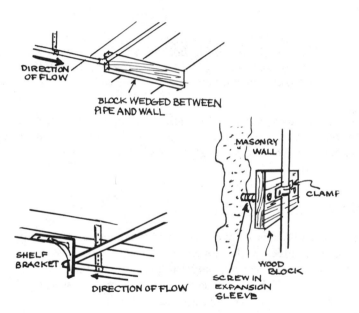

DIRECTION OF FLOW

BLOCK WEDGED BETWEEN PIPE AND WALL

MASONRY WALL

CLAMP

SHELF BRACKET

DIRECTION OF FLOW

SCREW IN EXPANSION SLEEVE

WOOD BLOCK

FIGURE 10-19
Water hammer can be caused by pipes that are not secured tightly. Shown here are ways to secure such pipes.

Water hammer may be prevented or its severity reduced by installing an air chamber just ahead of the faucet. The air chamber may be a piece of air-filled pipe or tubing, about 2 feet long, extending vertically from the pipe. It must be airtight. Commercial devices designed to prevent water hammer are also available.

An air chamber requires occasional replenishing of the air to keep it from becoming waterlogged—that is, full of water instead of air.

A properly operating hydropneumatic tank, such as the type used in individual water systems, serves as an air chamber, preventing or reducing water hammer.

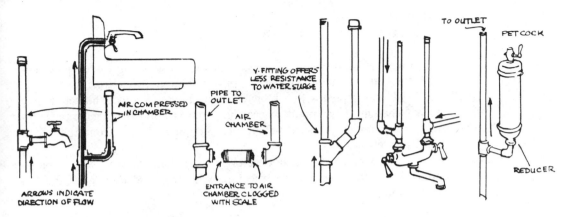

FIGURE 10-20

To eliminate knocking, you can install air chambers to relieve pressure in the line. If you already have air chambers, make certain that they are not clogged with scale.

FROZEN WATER PIPES

In cold weather, water may freeze in underground pipes laid above the frost line, in pipes in unheated buildings, in open crawl spaces under buildings, or in outside walls.

When water freezes, it expands. Unless a pipe can also expand, it may rupture when the water freezes. Iron pipe and steel pipe do not expand appreciably. Copper pipe stretches to some degree, but does not resume its original dimensions when thawed out; repeated freezings will cause it to fail eventually. Flexible plastic tubing can stand repeated freezes, but it is good practice to prevent it from freezing.

preventing freezing

Pipes may be insulated to prevent freezing, but this is not a completely dependable method. Insulation does not stop the loss of heat from the pipe —it merely slows it down—and the water may freeze if it stands in the pipe long enough at below-freezing temperature. Also, if the insulation becomes wet, it may lose its effectiveness.

Electric heating cable can be used to prevent pipes from freezing. The cable should be wrapped around the pipe and covered with insulation.

thawing

Use of electric heating cable is a good method of thawing frozen pipe, because the entire heated length of the pipe is thawed at one time.

Thawing pipe with a blowtorch can be dangerous. The water may get hot enough at the point where the torch is applied to generate sufficient steam under pressure to rupture the pipe. Steam from the break could severely scald you.

Thawing pipe with hot water is safer than thawing with a blowtorch. One method is to cover the pipe with rags and then pour the hot water over the rags.

FIGURE 10-21

The safest way to thaw frozen pipes is to wrap them with rags and pour on hot water, working from the faucets back. A propane torch can also be used, but care must be taken here. Faucets should be open to let off steam pressure.

When thawing pipe with a blowtorch, hot water, or similar methods, open a fauet and start thawing at that point. The open faucet will permit steam to escape, thus reducing the chance of the buildup of dangerous pressure. Do not allow the steam to condense and refreeze before it reaches the faucet.

Underground metal pipes can be thawed by passing a low-voltage electric current through them. The current will heat the entire length of pipe through which it passes. Both ends of the pipe must be open to prevent the buildup of steam pressure.

Caution: This method of thawing frozen pipe can be dangerous. It cannot be used to thaw plastic tubing or other non-electricity-conducting pipe or tubing.

REPAIRING WATER CLOSETS

Water closets vary in general design and in the design of the flushing mechanism. They are, however, enough alike that general repair instructions can be given for all designs.

flushing mechanism

Figure 10-22 shows a common type of flushing mechanism. Parts that usually require repair are the flush valve, the intake (float) valve, and the float ball.

In areas of corrosive water, the usual copper flushing mechanism may deteriorate in a comparatively short time. In such cases, it may be advisable to replace the corroded parts with plastic parts. You can even buy plastic float balls.

flush valve

The rubber ball of the flush valve may get soft or out of shape and fail to seat properly. This causes the valve to leak. Unscrew the ball from the lift wire and install a new one.

The trip lever or lift wire may corrode and fail to work smoothly, or the lift wire may bind in the guides. Disassemble and clean off corrosion or replace parts as necessary.

Most plumbing codes require a cutoff valve in the supply line to the flush tank, which makes it unnecessary to close down the whole system. If this valve was not installed, you can stop the flow of water by lifting and propping up the float with a piece of wood. Be careful not to bend the float rod out of alignment.

intake (float) valve

A worn plunger washer in the supply valve will cause the valve to leak. To replace the washer:

1.   Shut off the water and drain the tank.
2.   Unscrew the two thumbscrews that hold the levers and push out the levers.
3.   Lift out the plunger, unscrew the cup on the bottom, and insert a new washer. The washer is made of material such as rubber or leather.
4.   Examine the washer seat. If nicked or rough, it may need refacing. If the float-valve assembly is badly corroded, replace it.

float ball

The float ball may develop a leak and fail to rise to the proper position. (Correct water level is about 1 inch below the top of the overflow

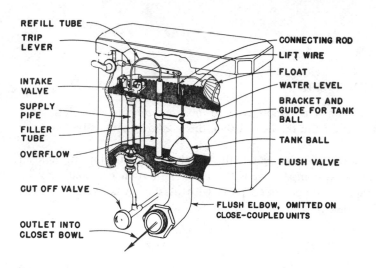

REFILL TUBE
TRIP LEVER
INTAKE VALVE
SUPPLY PIPE
FILLER TUBE
OVERFLOW
CUT OFF VALVE
OUTLET INTO CLOSET BOWL

CONNECTING ROD
LIFT WIRE
FLOAT
WATER LEVEL
BRACKET AND GUIDE FOR TANK BALL
TANK BALL
FLUSH VALVE
FLUSH ELBOW, OMITTED ON CLOSE-COUPLED UNITS

FIGURE 10-22

Parts of a typical water closet. If the flushing mechanism is in very bad shape, it is easy to remove by unscrewing it from the tank.

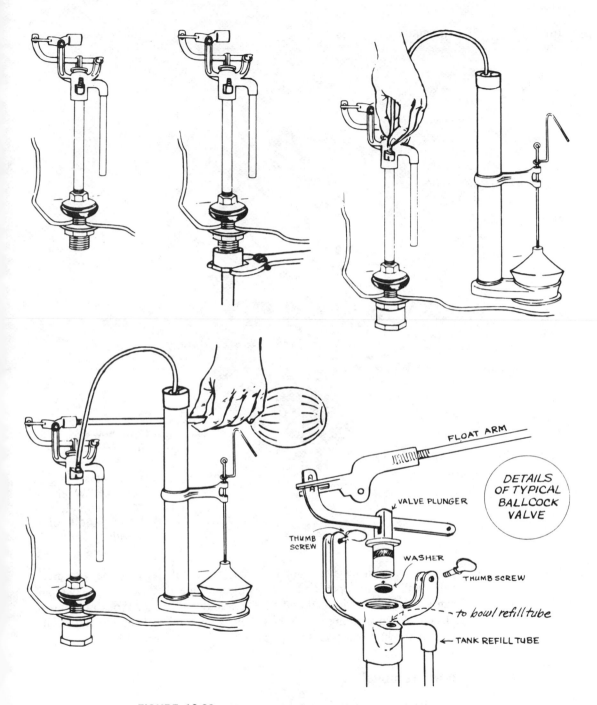

FLOAT ARM

DETAILS
OF TYPICAL
BALLCOCK
VALVE

VALVE PLUNGER

THUMB
SCREW

WASHER

THUMB SCREW

to bowl refill tube

TANK REFILL TUBE

FIGURE 10-23
The text details various ways to handle troubles in the water closet
mechanism. If necessary, this can be unscrewed from underneath and
replaced with a new one.

161

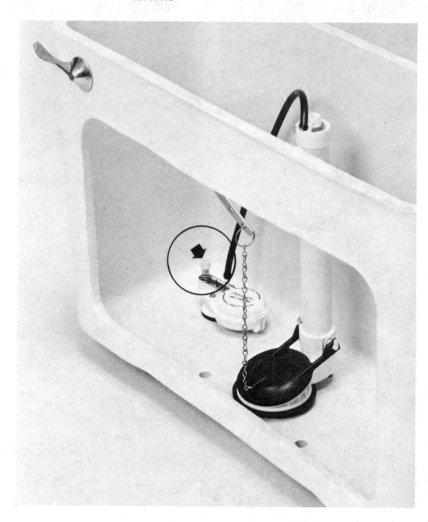

FIGURE 10-24
This new type of flushing mechanism eliminates the ballcock. Water flow can be adjusted by turning the knob detailed.

tube, or enough to give a good flush.) If the ball fails to rise, the intake valve will remain open and water will continue to flow. Brass float balls can sometimes be drained and the leak soldered. Other types must be replaced. When working on the float ball, be careful to keep the rod aligned so that the ball will float freely and close the valve properly.

bowl removal

An obstruction in the water-closet trap or leakage around the bottom of the water-closet bowl may require removal of the bowl, though obstructions can usually be cleared with a plunger. Follow this procedure for removing the bowl:

1. Shut off the water.
2. Empty the tank and bowl by siphoning or sponging out the water.
3. Disconnect the water pipes to the tank.
4. Disconnect the tank from the bowl if the water closet is a two-piece unit. Set the tank where it cannot be damaged. Handle the tank and bowl carefully; they are made of vitreous china or porcelain and are easily chipped or broken.
5. Remove the seat and cover from the bowl.
6. Carefully pry loose the bolt covers and remove the bolts holding the bowl to the floor flange. Jar the bowl enough to break the seal at the bottom. Set the bowl upside down on something that will not chip or break it.
7. Remove the obstruction from the discharge opening.
8. Place a new wax seal around the bowl horn and press it into place.
9. Set the bowl in place and press it down firmly. Install the bolts

FIGURE 10-25

When the toilet is clogged, it usually can be fixed by plunging. The cup should be covered with water (fill the bowl almost to the brim) while doing this.

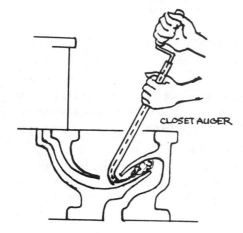

CLOSET AUGER

FIGURE 10-26

If the plunger doesn't remove the clog, a closet auger usually will. If this doesn't work, it means the obstruction is further down in the pipe, and you will need to use a long snake to remove it.

that hold it to the floor flange. Draw the bolts up snugly, but not too tightly, because the bowl may break. The bowl must be level; keep a level on it while drawing up the bolts. If the house has settled, leaving the floor sloping, it may be necessary to use shims to make the bowl set level. Replace the bolt covers.

10. Install the tank and connect the water pipes to it. It is advisable to replace all gaskets, after first cleaning the surfaces thoroughly.

11. Test for leaks by flushing a few times.

12. Install the seat and cover.

tank "sweating"

When cold water enters a water closet tank, it may chill the tank enough to cause "sweating" (condensation of atmospheric moisture on the outer surface of the tank). This can be prevented by insulating the tank to keep the temperature of the outer surface above the dew point temperature of surrounding air. Insulating jackets or liners that fit inside water-closet tanks and serve to keep the outer surface warm are available from plumbing-supply dealers.

CLEANING CLOGGED DRAINS

Drains may become clogged by objects dropped into them or by accumulations of grease, dirt, or other matter.

If the obstruction is in a fixture trap, usually the trap can be removed and cleared. If the obstruction is elsewhere in the pipe, other means must be used.

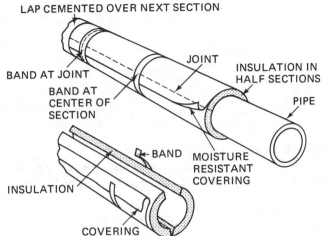

LAP CEMENTED OVER NEXT SECTION

BAND AT JOINT

BAND AT CENTER OF SECTION

JOINT

INSULATION IN HALF SECTIONS

PIPE

INSULATION

BAND

MOISTURE RESISTANT COVERING

COVERING

FIGURE 10-27

If pipes sweat, wrapping them with insulation will be somewhat helpful.

Cleanout augers—long, flexible, steel cables commonly called "snakes"—may be run down drain pipes to break up obstructions or to hook onto and pull out objects. Augers, including electrically powered ones, are made in various lengths and diameters and are available at hardware and plumbing-supply stores.

Small obstructions can sometimes be forced down or drawn up by use of an ordinary rubber force cup (plunger or "plumber's friend").

Grease and soap clinging to a pipe can sometimes be removed by flushing with hot water for 10 minutes or so. The use of chemical cleaners is not recommended. They may not do anything for a solid blockage, and if they don't work, you will have to plunge or use a snake, exposing yourself to the water with the caustic in it.

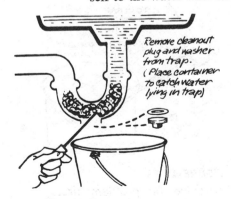

Remove cleanout plug and washer from trap. (Place container to catch water lying in trap)

FIGURE 10-28

If the plunger doesn't remove the clog, take off the trap plug or the trap itself and clean it as shown.

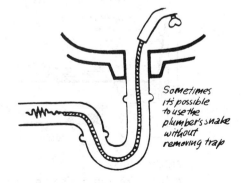

Sometimes it's possible to use the plumber's snake without removing trap

FIGURE 10-29

On some drains you will have to use a snake; this can be used without removing the trap.

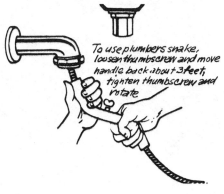

To use plumbers snake,
loosen thumbscrew and move
handle back about 3 feet,
tighten thumbscrew and
rotate

FIGURE 10-30

On other drains, you will have to take the trap off and
insert the snake into the pipe coming out of the wall.

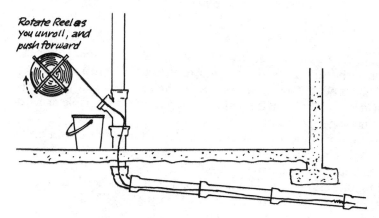

Rotate Reel as
you unroll, and
push forward

FIGURE 10-31

On a clogged toilet or
lavatory, the blockage may
be in the building drain. To
remedy this, first loosen
the cleanout plug in the
basement enough to let
water and waste flow into a
container. Then insert a
coiled cleanout tape into
the cleanout and feed it in
until the obstruction is
cleared. For the plumber,
an electric tape is best.

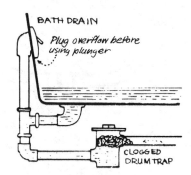

BATH DRAIN

Plug overflow before
using plunger

CLOGGED
DRUM TRAP

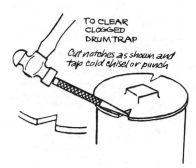

TO CLEAR
CLOGGED
DRUM TRAP

Cut notches as shown and
tap cold chisel or punch

FIGURE 10-32

A bath drain can also be
plunged, but the problem
might be a clogged drum
trap.

FIGURE 10-33

To solve this problem, first
remove the trap cover as
shown and insert the snake.

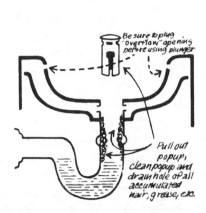

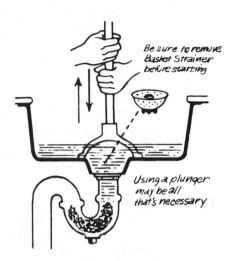

FIGURE 10-34
Lavatory drains often get
clogged with hair and lint.

FIGURE 10-35
A plunger is also good for
clearing lavatory drains.

Sand, dirt, or lint sometimes clog floor drains. Remove the strainer and ladle out as much of the sediment as possible. You may have to carefully chip away the concrete around the strainer to free it. Flush the drain with clean water. If pressure is needed, use a garden hose. Wrap cloths around the hose where it enters the drain to prevent backflow of water. You may have to stand on this plug to keep it in place when the water is turned on.

Occasional flushing of floor drains may prevent clogging.

Garden hoses, augurs, rubber force cups and other tools used in direct contact with sewage are subject to contamination. Do not later use them for work on your potable water supply system unless they have been properly sterilized.

OUTSIDE DRAINS

Roots growing through cracks or defective joints sometimes clog outside drains or sewers. You can clear the stoppage temporarily by using a root-cutting tool. However, to prevent future trouble, you should re-lay the defective portion of the line, using sound pipe and making sure that all joints are watertight.

If possible, sewer lines should be laid out of the reach of roots. But if this is impossible or impracticable, consider using impregnated fiber pipe, which tends to repel roots.

DRAINING PLUMBING IN A VACANT HOUSE

Shut off the water supply at the main shutoff valve at the street; then, beginning with those on the top floor, open all faucets, and leave them open. When water stops running from these faucets open the cap in the main pipe valve in the basement, and drain the remaining water into a pail or tub. This cap must be closed until the faucets have run dry, or the house water supply will flow from this valve and flood the basement.

Remove all water in the traps under sinks, water closets, bathtubs, and lavatories by opening the cleanout plugs at the bottom of traps and draining them into a pail. If no traps are provided, use a force pump or other method to siphon the water out. Sponge all the water out of the water closet bowls. Clean out all water in the flush tank.

Fill all traps with kerosene, crude glycerine, or one of the non-freeze compounds used in automobiles. This will seal off bad odors from waste pipes. Alcohol and kerosene are a recommended mixture, as the kerosene will rise to the top and prevent evaporation of the alcohol. Two quarts of this mixture will be needed for each toilet trap; other plumbing fixtures need less. Do not forget to fill the trap in the flood drain in the basement.

Drain all hot-water tanks. Most water tanks are equipped with a vented tube at the top which lets air in and allows the water to drain out the faucet at the bottom. Make sure that all horizontal pipes drain properly. Air pressure will get rid of trapped water in these pipes, but occasionally the piping may have to be disconnected and drained.

If the house is heated by hot water or steam, drain the heating pipes and boiler before leaving. Fires should be completely out and the main water supply turned off at the basement wall or street. Draw off the water from the boiler by opening the draw-off valve at the lowest point in the system.

Open the water supply valve to the boiler so that no water will be trapped above it. If you have a hot water system, begin with the highest radiators and open the air valve on each as fast as the water lowers. Every radiator valve must be opened on a one-pipe system to release condensation. Refill your heating system before lighting the boiler.

If you have ever had problems with mathematics, it is probably because math was a bore. You had no use for it, so it made little sense. This math will make a lot of sense because it will make dollars for you. We are going to include in it only those things you need for the job as you first start out. You can do plumbing at many levels. As a helper, you still have to know pipe sizes, how long a nipple is, or how many elbows you need. You will need a little figuring on pipe lengths to allow for the pipe threading into a fitting. As a plumber, you will want to know how many fixtures will be required and how much water they will use up. If you cannot find the proper pipe sizes to bring in the water without losing pressure, the contractor will soon find himself a new plumber who can.

A plumbing contractor uses math all the time. He must know costs, the discounts that he can get, the taxes, deductions, and salaries he pays out, and he must figure time and material on the job. If he can do those computations, he can also compute his *profit*.

You may have been a whiz, or a real stone in your last math course, but we are going to make this math refresher course really easy by starting from scratch. However, instead of boring you to tears with it during eight grades, we are going to get through it in one fast sweep.

NUMBERS

Whole numbers are the easiest to deal with. You count by ones. Children start out by counting ones on their fingers. The old Romans did not do much better—

11

Mathematics
Down the
Drain

they wrote I for 1 and II for 2. We use *digits* for all numbers from 1 through 9, that is, 1-2-3-4-5-6-7-8-9. Zero or 0 means that there are no ones. When we get to ten, we write two digits: 10. This means one ten and no ones. 37 is 3 tens and 7 ones. 203 means 2 hundreds, no tens, and 3 ones. It is very handy, because you know the number at the right is units— or ones—the second number is tens, and the third number from the right is hundreds.

Think of the first digit at the right of the number as a bucket—the most that it can hold is 9 (nine) ones. As soon as the bucket is full, and you try to add an extra one, it empties into the next column and becomes a ten. For instance, you add 9 + 1 and it spills over to the next column and becomes 10. Add 19 + 4 and you get 23 which is 2 tens and 3 ones. When you get up to 99, adding another one makes the number spill over to the third column—hundred—100. As before, you have ones all the way at the right, then tens, then hundreds, then thousands.

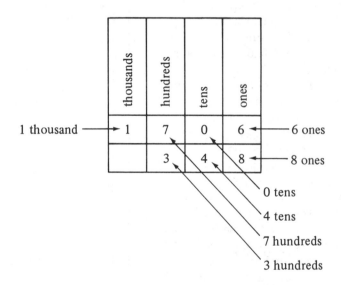

You can tell the sergeant's rank by the number of stripes he has. By the same token, you can tell the rank of a number by the place it has. The *ones* are all the way at the right. Next come *tens, hundreds, thousands, ten thousands,* etc.

The nice thing about it is that each rank is ten times bigger than the last one as you go from right to left. Ten tens (10 × 10) is a hundred (100). Ten hundreds is a thousand (10 × 100 = 1,000). Just to make it easier to read numbers, we put commas after each three digits, such as 1,956 or 2,987,345.

millions	hundred thousands	ten thousands	thousands	hundreds	tens	ones
1	9	7	8	5	4	3
9	0	8	2	6	3	7

ADDITION

When you see a + sign it means "plus," or "add." 5 + 5 is ten. To save writing you use the = (equals) sign to show what the answer will be. For instance, 6 + 7 = 13.

You may laugh, but the trouble most people have with math is that they did not remember the very simple things in the first two grades, and it has hindered them ever since. Adding digits is something you have to memorize, just like you remember your name. If the girl at the checkout counter can do it, so can you.

$$1 + 1 = 2 \quad 1 + 2 = 3 \quad 1 + 3 = 4 \quad 1 + 4 = 5 \quad 1 + 5 = 6$$
$$1 + 6 = 7 \quad 1 + 7 = 8 \quad 1 + 8 = 9 \quad 1 + 9 = 10$$

$$2 + 1 = 3 \quad 2 + 2 = 4 \quad 2 + 3 = 5 \quad 2 + 4 = 6 \quad 2 + 5 = 7$$
$$2 + 6 = 8 \quad 2 + 7 = 9 \quad 2 + 8 = 10 \quad 2 + 9 = 11$$

$$3 + 1 = 4 \quad 3 + 2 = 5 \quad 3 + 3 = 6 \quad 3 + 4 = 7 \quad 3 + 5 = 8$$
$$3 + 6 = 9 \quad 3 + 7 = 10 \quad 3 + 8 = 11 \quad 3 + 9 = 12$$

You can complete your own table, which will help you memorize it. If you have to "think" of the answer, you have not memorized well enough.

When you add larger single digits, like 9 + 9, there is not enough room in the ones column, and the number spills over to the tens; for instance, 9 + 9 = 18. If you add a column of numbers, more than one ten may spill over. For instance:

$$
\begin{array}{r}
③ \\
1 \\
9 \\
7 \\
6 \\
8 \\
\hline
31
\end{array}
$$

Adding numbers with several digits each is simple, if you simply keep the columns neatly lined up. Most errors occur only because the different columns are jumbled. For instance, add the following:

		1	1	1	
		1	2	6	7
			3	2	1
+	9	4	7	6	4	8
	9	4	9	2	3	6

$$7 + 1 + 8 = 16$$

Write down the 6 and transfer the 1, which represents tens in the ten column. The 1 you transferred heads the ten column.

$$1 + 6 + 2 + 4 = 13$$

The 13 is really 13 tens, since they all came from the ten column. Write 3 in the ten column and transfer the 10 into the next column, as 1, or one hundred.

$$1 + 2 + 3 + 6 = 12$$

This 12 is really 12 hundreds, so write the 2 in the sum section of the hundreds column and transfer the 10 into the next column as a 1, or one thousand. This 1 now heads or tops the thousands column.

$$1 + 1 + 7 = 9$$

The 9 is nine thousand; this is the sum of the thousands column.
4 is the ten-thousands column—there was nothing to add.
9 is the hundred-thousands column—again, there was nothing to add. The sum of 949,236 is obtained, and the comma is placed after the first three digits on the right. You read the sum (total) number as nine hundred forty-nine thousand, two hundred thirty-six.

SUBTRACTION

If math comes hard to you, you will find that a couple of rolls of pennies is all that it costs to learn subtraction. From a stack of 20 pennies, take away

8, count what is left, and memorize the result. Keep at it for a few days and you will be as fast as the next person.

Subtracting from a number where some of the digits are zero calls for a little extra thinking.

You have no ones and no tens, so you borrow a hundred. This transfers 100 or 10 tens into the tens column. Of those 10 tens you borrow one, $10 - 1 = 9$ (in the tens column.) The actual values are written in for your convenience here.

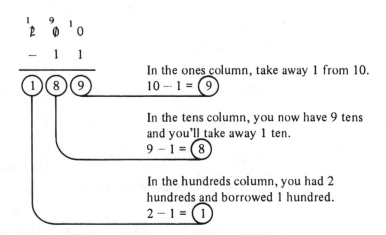

In the ones column, take away 1 from 10. $10 - 1 = 9$

In the tens column, you now have 9 tens and you'll take away 1 ten. $9 - 1 = 8$

In the hundreds column, you had 2 hundreds and borrowed 1 hundred. $2 - 1 = 1$

When your boss takes a deduction from your pay for taxes, you have just understood the meaning of subtraction. Subtraction is taking away a smaller number from a larger one.

You have a 9-foot bathroom with a 6-foot vanity; there will be only 3 feet left for the water closet.

Subtraction is the reverse of addition. You added $7 + 9 = 16$. In subtraction $16 - 9 = 7$.

In subtracting $19 - 2 = 17$, you had enough ones to complete the subtraction without borrowing from the tens. In $27 - 8$ there are not enough ones to complete the operation.

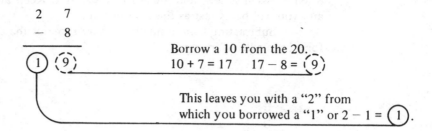

In an actual subtraction you shortcut all this.

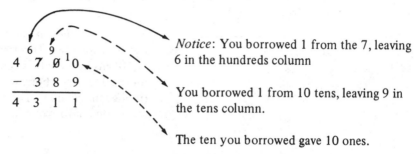

Notice: You borrowed 1 from the 7, leaving 6 in the hundreds column

You borrowed 1 from 10 tens, leaving 9 in the tens column.

The ten you borrowed gave 10 ones.

proof of subtraction

Add the number you are subtracting (the smaller number) to the answer. The result is always the big number. If it is not, there is a mistake.

$$
\begin{array}{rrrr}
4700 & 389 & 200 & 11 \\
-389 & +4311 & -11 & +189 \\
\hline
4311 & 4700 & 189 & 200 \\
\end{array}
$$

You have just become a master subtracter.

MULTIPLICATION

Multiplication is used all the time. You buy six fittings at $2.00 apiece. You can add $2.00 six times and come up with the right answer ($12.00), or you can multiply and get $6 \times 2 = 12$. Figuring heating loads calls for finding floor areas in square feet. To find an area, you simply take the width and length of a room and multiply. For instance, a room that measures 12 by 18 measures 216 square feet. Now let us get into details.

The basic brick for multiplication is addition. Four times three simply means to take four threes and add them together.

$$4 \times 3 = 3 + 3 + 3 + 3 = 12$$

All multiplication problems boil down to memorizing a table of multiplication. The only way you will do it is through simple practice. You cannot learn how to run or climb trees by reading a book. By the same token, you cannot learn to multiply without memorizing the facts. Better yet, you can create your own multiplication table by simple addition. Here is how:

$3 \times 0 = 0$ Three times nothing is still nothing.
$3 \times 1 = 3$
$3 \times 2 = 6$
$3 \times 3 = 9$
$3 \times 4 = 12$

Table 11-1 is a table of multiplication (up to 10×10). You use it by finding the number you want to multiply along one side and then moving across the row until you find the column that contains the number by which you are multiplying. It is just like reading a map. The answer is where the row and the column meet.

Table 11-1 TABLE OF MULTIPLICATION

1	2	3	4	5	6	7	8	9	10
2	4	6	8	10	12	14	16	18	20
3	6	9	12	15	18	21	24	27	30
4	8	12	16	20	24	28	32	36	40
5	10	15	20	25	30	35	40	45	50
6	12	18	24	30	36	42	48	54	60
7	14	21	28	35	42	49	56	63	70
8	16	24	32	40	48	56	64	72	80
9	18	27	36	45	54	63	72	81	90
10	20	30	40	50	60	70	80	90	100

```
 327          8 × 7 = 56
 ×8           Write down the 6 and carry the 5.


  5
 327          8 × 2 = 16
 ×8           Now add the 5 you carried.
  6           16 + 5 = 21
              Write down the 1 and carry the 2.


 25
 327          8 × 3 = 24
 ×8           Add the 2 you were carrying.
 16           24 + 2 = 26


 327          Rule:   On the last operation:
 ×8           Do not carry—just write down the
2616          answer.
```

brief reminder

```
 327          You started out multiplying ones.
 ×8           8 × 7 = 56
  6           You enter the ones, but carry a 5,
              which is really a 50.
              In the next step you multiply 8 × 2;
              you are really multiplying 8 × 20 =
              160.

              6 stands for tens. However, you are
              carrying 5 more tens (50).

                 160
                +50


  2
 327          Enter the 21 ten and carry the 200.
 ×8
 16           8 × 3 is really 8 × 300, or 2400.
              Add the 200 you already have.

 327             2400
 ×8             +200
2616            2600
```

multiplying numbers with several digits

			Carry
749	$7 \times 9 = 63$. Write down 3		6
$\times 27$	and carry 6.		
5243	$7 \times 4 = 28$; $28 + 6 = 34$.		3
	Write down 4 and carry 3.		
	$7 \times 7 = 49$; $49 + 3 = 52$.		

You have multiplied 749 by 7. Next you have to multiply 749 by 20 and then add the two results. In actual practice you multiply by "2." Then all the numbers are offset by one space.

Offsetting by one space to the left automatically multiplies them by 10.

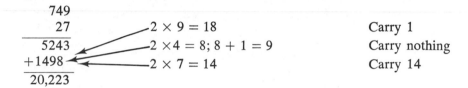

$2 \times 9 = 18$	Carry 1
$2 \times 4 = 8$; $8 + 1 = 9$	Carry nothing
$2 \times 7 = 14$	Carry 14

```
  749
   27
 5243
+1498
20,223
```

DIVISION

To divide is to find out how many times one number fits into another.

$$10 \div 2 = 5$$

$$||||| \quad |||||$$
$$\ \ 5 \qquad \ 5$$

Division is largely a table of multiplication in reverse.

$28 \div 7 = 4$

7 went into 28 4 times. To prove it yourself, retrace your steps. $7 \times 4 = 28$.

$2\overline{)842}$

$\quad\ 4$
$2\overline{)842}$

You begin by dividing the biggest units. At the left, $8 \div 2 = 4$.

$$\begin{array}{r} 42 \\ 2\overline{)842} \end{array}$$

Move to the next unit. $4 \div 2 = 2$.

$$\begin{array}{r} 421 \\ 2\overline{)842} \end{array}$$

Move to the next unit. $2 \div 2 = 1$.

Remember, for each digit in the number divided, you should have a digit in the answer.

To prove it to yourself, multiply your answer by the divisor, and you should get the original number that you were dividing.

$$\begin{array}{r} 421 \\ \times 2 \\ \hline 842 \end{array}$$

So far you have divided, in succession, the hundreds, the tens, and the ones. Let us go a step further.

$4\overline{)19,762}$

4 will not go into 1, so we shall divide a unit that will hold it. $19 \div 4 = 4$ with a remainder of 3.

$$\begin{array}{r} 4,940\frac{1}{2} \\ 4\overline{)19,762} \\ -16 \\ \hline 37 \\ -36 \\ \hline 16 \\ -16 \\ \hline 02 \end{array}$$

$4 \times 4 = 16$; $16 + 3 = 19$.
Multiply $4 \times 4 = 16$ and subtract it from 19.
Bring down the next digit: 7.
$37 \div 4 = 9$; $9 \times 4 = 36$; subtract it from 37.
Bring down the next digit: 6. $16 \div 4 = 4$.
There is no remainder: 0. You still bring down the next digit. $2 \div 4 = \frac{2}{4} = \frac{1}{2}$.

$$\begin{array}{r} 4940\frac{1}{2} \\ \times 4 \\ \hline 2 \\ 19760 \\ \hline 19762 \end{array}$$
$4 \times \frac{1}{2} = 2$

Again, a check will be to multiply.

If you have multiplied right, your answer should check.

FRACTIONS

You need fractions every time you pick up a ruler. Just measuring to the nearest inch is not good enough, so you have to find out how to handle parts of an inch, such as halves, quarters, or thirty-seconds. Every time you talk of pennies, like $2.97, you also deal in fractions, fractions of a dollar. Every time you divide one number by another, you have a fraction. For example, $^{27}/_4$, $^{27}/_5$, $^1/_2$, $^7/_{16}$ are all fractions.

The number at the top is the one you are dividing. It is called the *numerator*. The number at the bottom is doing the dividing. It is called the *denominator*. You will notice that $^{27}/_4$ is more than one. As a matter of fact, $^{27}/_4 = 6^3/_4$. $^{27}/_4$ is called an *improper fraction*—a fraction larger than one. You use improper fractions all the time, in adding fractions, which together equal more than one. For instance, $^3/_4 + ^3/_4 = ^6/_4 = 1^1/_2$.

common fractions and decimal fractions

Both common and decimal fractions are fractions. However, decimal fractions are special, in that the number doing the dividing, or the denominator, is a multiple of 10. In other words, you divide by 10, 100, 1000, and so on. You can write nine-tenths as $^9/_{10}$. In that form it is called a common fraction. Now write it as 0.9. It is still pronounced "nine-tenths," but it is written in decimal form. 14.64 is pronounced 14 and $^{64}/_{100}$—fourteen and sixty-four hundredths.

adding common fractions

A long time ago someone divided the inch into halves, quarters, eighths, sixteenths, thirty-seconds and sixty-fourths: $^1/_2$, $^1/_4$, $^1/_8$, $^1/_{16}$, $^1/_{32}$, $^1/_{64}$. We have been living with it ever since.

Dividing into halves means splitting the inch into two equal parts. Any half-inch anywhere on the ruler is of the same length. Two halves together make an inch.

$$\frac{1}{2} + \frac{1}{2} = 1 \qquad \text{You could write this } \frac{1+1}{2} = \frac{2}{2} = 1.$$

If it looks too simple, do not skip it, because all the other fractions are added the same way.

$$\frac{1}{2} + \frac{1}{2} + \frac{1}{2} = \frac{1+1+1}{2} = \frac{3}{2} \qquad \frac{2}{2} = 1, \text{ so} \frac{3}{2} = 1\frac{1}{2}$$

Quarters simply means splitting an inch into four equal parts.

$$^1/_4 + ^1/_4 = ^2/_4 \qquad ^1/_4 + ^1/_4 + ^1/_4 = ^3/_4$$

adding quarters and halves

Quarters can be added to each other.

$$\tfrac{3}{4} + \tfrac{3}{4} = \tfrac{6}{4}$$

Obviously, six quarters is more than one. $\tfrac{4}{4} = 1$, so $\tfrac{6}{4} = 1\tfrac{2}{4}$ or $1\tfrac{1}{2}$.

When adding mixed numbers, add the whole numbers first; then add the fractions and combine the result.

$$1\tfrac{3}{4} + 2\tfrac{1}{4}$$

$$1 + 2 = 3$$
$$3 + 1 = 4$$
$$\tfrac{3}{4} + \tfrac{1}{4} = \tfrac{4}{4} = 1$$

To add $\tfrac{1}{4} + \tfrac{1}{2}$, you first have to convert the half into quarters. Go into a store with a half a dollar, and you get two quarters in change. Same in arithmetic. $\tfrac{1}{4} + \tfrac{2}{4} = \tfrac{3}{4}$.

To add $\tfrac{1}{2} + \tfrac{1}{2} + \tfrac{1}{4} + \tfrac{3}{4}$, convert the $\tfrac{1}{2}$ into quarters.

$$\tfrac{2}{4} + \tfrac{2}{4} + \tfrac{1}{4} + \tfrac{3}{4} = \tfrac{8}{4} = 2$$

Eighths are handled in the same way. For example, an inch may be divided into eight equal parts. $\tfrac{1}{8}$ is simply one part out of eight. $\tfrac{2}{8}$ is 2 parts out of eight. $\tfrac{4}{8}$ is 4 parts out of eight.

Look at a ruler divided into 8 parts.

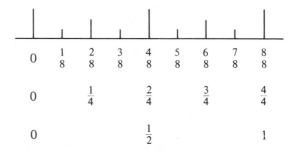

If you come to a mark that shows $\tfrac{4}{8}$, it does not take you long to recognize that it is also $\tfrac{1}{2}$ or $\tfrac{2}{4}$. $\tfrac{6}{8}$ is also $\tfrac{3}{4}$, and $\tfrac{8}{8}$ is one inch. *Any number divided by itself is one*—as $\tfrac{8}{8}$ is 1. Zero is the only exception—$\tfrac{0}{4}$ is still 0, and $\tfrac{0}{0}$ has no meaning.

You can add eighths to halves and quarters, but they all must be converted to eighths.

$$\tfrac{1}{8} + \tfrac{1}{2} + \tfrac{3}{4} = \tfrac{1}{8} + \tfrac{4}{8} + \tfrac{6}{8} = \tfrac{11}{8} = 1\tfrac{3}{8}$$

If you do not believe us, look up the values on the scale. By now you are getting the message. Fractions can be converted. *Rule:* You must multiply the top and the bottom of a fraction by the same amount.

$$\frac{1}{2} = \frac{1 \times 8}{2 \times 8} = \frac{8}{16} \qquad \frac{3}{4} = \frac{3 \times 4}{4 \times 4} = \frac{12}{16}$$

Rule: You can reduce fractions by dividing the top and the bottom by the same amount.

$$\frac{32}{64} = \frac{32/32}{64/32} = \frac{1}{2}$$

$$\frac{48}{64} = \frac{48/2}{64/2} = \frac{24}{32}$$

It is also equal to

$$\frac{48/4}{64/4} = \frac{12}{16} \text{ or } \frac{3}{4}$$

$$\frac{3}{16} + \frac{1}{2} = \frac{3}{16} + \frac{1 \times 8}{2 \times 8} = \frac{3}{16} + \frac{8}{16} = \frac{11}{16}$$

Convert the half to sixteenths—multiply the top and bottom by 8. *Memorize:*

$$\tfrac{1}{4} = \tfrac{2}{8} = \tfrac{4}{16} \qquad \tfrac{2}{4} = \tfrac{1}{2} = \tfrac{4}{8} = \tfrac{8}{16} = \tfrac{16}{32} \qquad \tfrac{3}{4} = \tfrac{6}{8} = \tfrac{12}{16} = \tfrac{24}{32}$$

subtracting fractions

The same conversion rules apply. All fractions must have the same common denominator—the same number at the bottom—before you can add or subtract them.

$$\frac{3}{4} - \frac{5}{16} =$$

Convert to sixteenths. 16 is 4 times bigger than 4, so we multiply ¾ by 4 top and bottom.

$$\frac{3 \times 4}{4 \times 4} = \frac{12}{16} \qquad \frac{12 - 5}{16} = \frac{7}{16}$$

Remember how we borrowed in subtraction of whole numbers? It is the same with fractions.

$$56\tfrac{1}{2} - 2\tfrac{3}{4} =$$

¾ is bigger than ½, so we also borrow a 1 from 56

$$56\tfrac{1}{2} = \quad 55 \quad 1\tfrac{1}{2}$$
$$-2\tfrac{3}{4} = \quad -2 \quad \tfrac{3}{4}$$

$$55 - 2 = 53$$

$$1\tfrac{1}{2} - \tfrac{3}{4} = \tfrac{6}{4} - \tfrac{3}{4} = \tfrac{3}{4} \text{ or } 53\tfrac{3}{4}$$

multiplying fractions

To multiply a fraction by a whole number, just multiply the numerator. For instance,

$$\frac{7}{16} \times 5 = \frac{7 \times 5}{16} = \frac{35}{16}$$

To reconvert, just divide 35 by 16.

$$
\begin{array}{r}
2 \\
16\overline{)35} \\
32 \\
\hline
3
\end{array}
$$

The answer is 2³⁄₁₆.

To multiply a fraction by another fraction, just multiply the two numerators (top numbers) and the two denominators (bottom numbers).

$$\frac{7}{32} \times \frac{9}{16} = \frac{7 \times 9}{32 \times 16} = \frac{63}{512}$$

$$
\begin{array}{r}
32 \\
\times 16 \\
\hline
192 \\
32 \\
\hline
512
\end{array}
$$

Remember: When fractions are multiplied, the answer will always be smaller than either fraction.

dividing fractions

To divide fractions, you can go through a lot of theory, or you can simply turn the fraction by which you are dividing upside down and proceed as though you are multiplying.

Reverse the last fraction.

$$\frac{1}{2} \div \frac{1}{4} = \frac{1}{2} \times \frac{4}{1} = \frac{1 \times 4}{2 \times 1} = \frac{4}{2} = 2$$

$$\frac{3}{4} \div \frac{7}{8} = \frac{3}{4} \times \frac{8}{7} = \frac{24}{28} = \frac{6}{7}$$

dividing a fraction by a whole number

Multiply the denominator by the whole number.

$$\frac{1}{2} \div 6 = \frac{1}{2 \times 6} = \frac{1}{12}$$

DECIMALS

In comparison to working with fractions, decimals are a hundred times easier, as long as you get over the first hurdles. With decimals you can crank all the questions into a pocket calculator and never make a mistake. However, before using your calculator, you have to understand decimals.

Rule: All numbers to the right of the decimal point represent a fraction. All numbers to the left of the decimal point represent the whole number. For instance, in 127.64, 127 is the whole number, and .64 is the fraction representing $^{64}/_{100}$. 0.64 means no ones and $^{64}/_{100}$.

Consider a number chart. You can tell at a glance the rank of each row of digits.

$$
\begin{array}{lcr}
10 & = & 10 \\
10 \times 10 & = & 100 \\
10 \times 10 \times 10 & = & 1000 \\
10 \times 10 \times 10 \times 10 & = & 10{,}000
\end{array}
$$

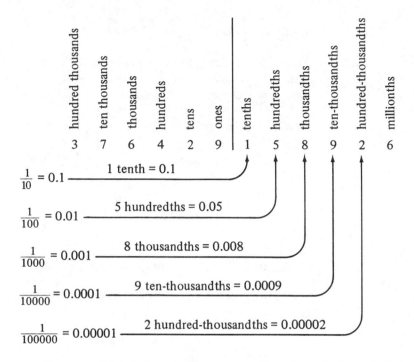

Just as with whole numbers, you tell rank and size of decimal numbers by the location of the digit—how far away it is from the decimal point.

reading decimals

Every time you read a micrometer or a set of calipers, you need decimals. These few examples will help you read them.

$.7 = \frac{7}{10}$; read as "7 tenths."

nine hundredths

$.99 = \frac{99}{100}$; read as "99 hundredths."

nine tenths, or 90 hundredths

$.08 = \frac{8}{100}$

no tenths   8 hundredths

$.002 = \frac{2}{1000}$; read as "two thousandths."

2 thousandths
no hundredths
no tenths

.025 = $^{25}/_{1000}$; read as "twenty-five thousandths"

5 thousandths
2 hundredths or 20 thousandths
no tenths

converting common fractions to decimals

Just divide the top, or numerator, by the bottom, or denominator. To do it by long division is madness when you can punch the buttons on the calculator. $^{27}/_{64}$ = .421875. You can also find decimal equivalents on the back of most good metal scales.

Converting decimals to fractions can be useful. For instance, you want to convert .769 to sixteenths. Multiply the decimal by 16. The answer is 12.304. This means a little over $^{12}/_{16}$. To convert the same .769 into thirty-seconds, multiply .769 by 32. The answer on the calculator is 24.608, or $^{25}/_{32}$, if we round off to the nearest whole number.

Remember and memorize some key decimals and their fractions.

DECIMALS	FRACTIONS
.125	$^{1}/_{8}$
.250	$^{1}/_{4}$
.375	$^{3}/_{8}$
.500	$^{1}/_{2}$
.625	$^{5}/_{8}$
.750	$^{3}/_{4}$
.875	$^{7}/_{8}$
.062	$^{1}/_{16}$
.031	$^{1}/_{32}$
.015	$^{1}/_{64}$

Usually thousandths or 3 place accuracy is enough when working with decimals.

multiplying by 10 and powers of 10

Multiplying by ten just calls for adding an extra zero at the end of the number. This moves all the digits up one rank (place) to the right. For instance;

$$1907 \times 10 = 19,070$$

Similarly,

$$2760 \times 10 = 27,600$$

Multiplying by 100 calls for adding two zeroes.

$$8764 \times 100 = 876,400$$

Multiplying by any power of 10 calls for adding a like amount of zeroes. For example, in multiplying any number $\times$ 1,000,000, there are six zeroes, so add 000,000 to the end of the number.

dividing by 10

Just move the decimal point one digit to the left when you divide by 10.

$$19,801 \div 10 = 1980.1$$

You have moved all digits down one rank.

dividing any power of 10

To divide by 100, move the decimal point 2 places to the left. To divide by 1,000, move the decimal 3 places to the left. To divide by 1,000,000, count the zeroes and move the decimal a like number of places.

multiplying decimals

Multiply as a normal multiplication. Count the number of decimals on each number. Add them and set the decimal point on your answer accordingly.

$$29.25 \times 2.15 = 66.8875$$

4 decimal places on the left of the equals sign = 4 decimal places on the right of the equals sign.

adding and subtracting decimals

Just line them up as in a conventional addition or subtraction. The decimal points must line up under each other. The decimal point of the answer comes down automatically.

$$\begin{array}{r} 375.160 \\ +39.0125 \\ \hline 414.1725 \end{array}$$

RATIOS

Ratio is another name for a fraction. ½ is the ratio of 1 to 2. 15 shovels of sand to one sack of cement is a ratio. ¾ is the ratio of 3 to 4. Scales for buildings are presented as ratios. "Scale: ¹⁄₁₆ in. = 1 ft." means ¹⁄₁₆ in. is equal to 1 ft. Thus, ³⁄₁₆ in. on the drawing will equal 3 feet.

AREAS

Lengths are measured in inches, feet, or yards. Areas are measured in anything from acres to square inches.

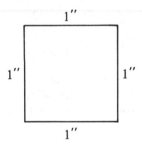

A square inch is simply what it says, or a square with a one-inch length and a one-inch width. By the same token, a square foot is a square with a one-foot length and a one-foot width. A square yard? You guessed it—a square one yard in length and one yard in width.

To measure an area you could fit a series of square inches into it as follows:

1	2	3	4	5	6
2					
3					
4					

There are four rows of six; 4 × 6 = 24. You could also say six columns of 4, or 6 × 4 = 24.

Measuring floor area in a room is too cumbersome in square inches, so you use a larger *unit,* such as square feet. In a room 12 feet by 15 feet, the area is $12 \times 15 = 180$ square feet.

FORMULAS

A formula is like a magic recipe. Crank in the right numbers for the letters and you get the answer.

area of a rectangle

$$A = L \times W$$

Area equals length times width. If the length is 9 feet and the width is 9½ feet, the area A is 9 feet $\times$ 9½ feet, or 85½ square feet.

area of a square

$A = s^2$ (pronounced "s squared")
s is the length of one side.
Area $= s$ times s (a square is a number multiplied by itself)

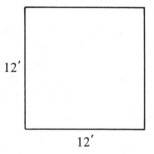

$A = 12$ ft. $\times$ 12 ft. $= 144$ square feet

area of a triangle

To measure the area of a triangle, you have to know the base and the height.

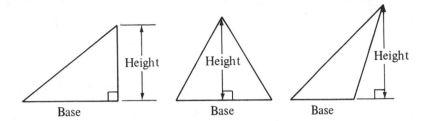

Base is the bottom and height is a distance measured at right angles to the base. The formula is

$$\text{Area of a triangle} = \tfrac{1}{2}\,\text{base} \times \text{height}$$

If you really want to prove it to yourself, fold a rectangle in two along a diagonal,

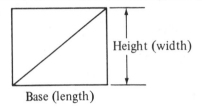

and you will see that a triangle is half the area of the rectangle. For instance, a triangle with a 6-foot base and a 10-foot height will have an area of

$$\tfrac{1}{2} \times 6 \times 10 = 30 \text{ square feet}$$

area of a trapezoid

A trapezoid is a triangle with the top cut off, or, to put it another way, a rectangle with triangles on each side.

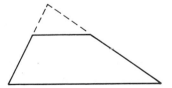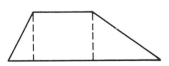

formula

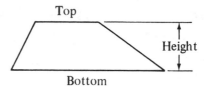

$$\text{Area of a trapezoid} = \frac{\text{top} + \text{bottom}}{2} \times \text{height}$$

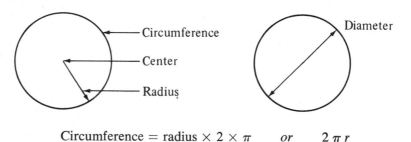

Top

Height

Bottom

area of a circle

If you think the area of a circle is not important, try finding the flow area and pipe sizes without it.

Circumference

Center

Radius

Diameter

$$\text{Circumference} = \text{radius} \times 2 \times \pi \qquad or \qquad 2\,\pi\,r$$

π (pi—pronounced "pie") is the Greek letter that stands for 3.14, or $22/7$. If you want to get fancy, $\pi = 3.1415926$.

$$\text{Circumference} = \pi \times \text{diameter} \qquad or \qquad \pi\,d$$

A circle 2 feet in diameter will have a 6.28-foot circumference.

$$\text{Area} = \quad \pi r^2 \quad \text{or} \quad \frac{\pi d^2}{4} \quad \text{or} \quad .7853d^2$$

A pipe with a .75-inch diameter inside offers a (.7853 × .75 × .75 = .44178) .44178-square-inch area, or $7/16$ square inch.

PER CENT

Per cent means part of a hundred. The symbol % means percentage. If a bank pays five per cent (5%) interest, it gives you five dollars for every hundred dollars you keep there. It is the bank's way of making it attractive for you to keep your money there instead of storing it under your mattress.

When you go to work as a plumber and are in the trade, doing business with a supplier, you will want a better price for the materials than the ordinary homeowner pays. The plumbing supply house obliges by offering you a discount. In turn, the plumbing supply house buys from a plumbing wholesaler. For them to make a profit, the plumbing wholesaler will offer the supply house a larger discount than the one that you get.

A manufacturer does not have enough men on staff to service the individual plumber or even all plumbing supply houses. He relies on individual wholesale distributors to contact all the plumbing supply houses and promote his product. The wholesale distributor expects an even larger discount from the manufacturer.

When you have an open account with a plumbing supply house and so do several hundred other plumbers, a large amount of money remains outstanding until bills are sent. The supply house would rather have its money available for merchandise instead of just outstanding, so they offer a cash discount, generally two per cent (2%) but sometimes as high as three (3%), to make it profitable for you to pay cash and to give them more operating money. If your bill is $329.75, a 2% discount will be figured as follows: 2% = 2 cents on a dollar = .02. $329.75 × .02 = $6.59. If you have the money to pay cash, you will save $6.59.

The manufacturer has a list price of $127.50 for an item; however, being in the trade, you get a 25 per cent discount. Your discount will be $127.50 × .25, or $31.87. Now you take your 2% discount, or $95.63 × .02 = $1.91.

$$
\begin{array}{ll}
\$127.50 & \$95.63 \\
-\ 31.87 & -\ 1.91 \\
\hline
\$\ 95.63 & \$93.72 \text{ is what you pay.}
\end{array}
$$

You can save an operation in figuring discount by subtracting the percentage from 100. For instance, a 30% discount means that you will be paying 70% of the total. On a $459.00 item with a 30% discount, your figures will be $459.00 × .70 = $321.30. Note that 70% is .70.

A succession of discounts is called a *chain discount*. If you are figuring a chain discount of 30% and 2%, you will not have a 32% discount. The 30% discount is figured first, leaving 70%. The 2% cash discount will make you pay 98% of the total.

$$70\% \times 98\% = 68.6\%$$

markup

Your markup is based on your cost of doing business and what your competitors are charging. If your accountant decides that a 50% markup is reasonable in the light of your operating expenses, your sales price, for any given item, will be the 100% you paid for it plus 50%, or 150%. If a water heater costs you $530.00, the 150% markup will bring it to $795.00.

back-siphonage

The flow of water into a water-supply system from any source other than its regular source. Also called back-flow.

back vent pipe

That part of a vent line which connects directly with an individual trap underneath or behind a fixture and extends to the branch or main, soil or waste pipe at any point higher than the fixture or fixture trap it serves. This is sometimes called an individual vent.

branch

A branch is any part of a piping system other than a main.

branch vent

A vent pipe connecting from a branch of the drainage system to a vent stack.

by-pass

Any method by which water may pass around a valve, fixture appliance connection, or length of pipe. Sometimes applied to an erroneous connection between a drain pipe and a vent pipe, which will allow sewer air to enter the building.

collar

A sleeve in back of a flange.

continuous waste

A waste from two or more fixtures connected to a single trap.

deep seal trap

A trap with a seal of 4 inches or more.

dip of a trap

The lowest portion of the inside top surface on the channel through the trap.

Glossary of Plumbing Terms

drain

A sewer or other pipe or conduit used for carrying ground water, surface water, storm water, waste water, or sewage.

drainage fitting

A fitting used on drainage pipes. A distinctive feature is the shoulder against which the connecting pipe rests so there is a smooth interior surface.

drum trap

A trap consisting of a cylinder with its axis vertical. The cylinder is larger in diameter than the inlet or outlet pipe; it is usually about 4 inches in diameter with 1½″ inlet and outlet pipes.

dry vent

A vent that does not carry water or water-borne wastes.

fittings

Parts of a pipe line other than straight pipe or valves, such as couplings, elbows, tees, unions and increasers.

fixture

A receptacle attached to a plumbing system other than a trap in which water or wastes may be collected or retained for discharge into the plumbing system.

fixture branch

The supply pipe between the fixture and the water distributing pipe.

fixture drain

The drain from the trap of a fixture to a junction of the drain with any other drain pipe.

flush bushing

A bushing without a shoulder that fits flush into the fitting with which it is to connect.

group vent

A branch vent that works for two or more traps.

horizontal branch

A branch drain extending laterally from a waste stack, with or without vertical sections or branches, that receives the discharge from one or more fixture drains and conducts it to the soil or waste stack or to the building drain.

house drain

That part of the lowest horizontal piping of a plumbing system which receives the discharge from soil, waste, and other drainage pipes inside of a building drain.

increaser

A coupling with one end larger than the other. Sometimes, more specifically, a pipe fitting to join the end of a small coupling with inside connection to the end of a larger pipe with outside connection.

local vent

A pipe or shaft serving to convey foul air from a plumbing fixture or a room to the outer air.

loop or circuit vent

A continuation of a horizontal soil or waste pipe beyond the connection at which liquid wastes from a fixture (or fixtures) enter the waste or soil pipe. The extension usually becomes vertical immediately beyond this connection. The base of the vertical portion of the vent may be connected to the horizontal portion of the soil or waste stack between fixtures. The Plumbing Manual and the Housing Code differentiate between a circuit vent and a loop vent. A loop vent loops back and connects with a soil or waste stack vent instead of a vent stack.

plug

A pipe fitting used for closing the opening in another fitting.

primary branch

A primary branch of the building drain is the single sloping drain from the base of a stack to its junction with the main building drain or with another branch.

relief vent

A branch from the vent stack, connected to a horizontal branch between the fixture branch and the soil or waste stack for circulation of air between the vent stack and the soil or waste stack.

self-siphonage

Breaking the seal of a trap as a result of removing the water by the discharge of the fixture to which the trap is connected.

side outlet ell

An ell with outlet at a right angle to the plane of the run.

side vent

A vent connection to the drain pipe through a 45 degree Y fitting.

siphonage

Suction created by the flow of liquids in pipes.

slip joint

A connection in which one pipe slides into another. The joint is made tight with approved gasket, packing or calking.

soil pipe

A pipe which carries liquid wastes and excrement.

soil stack

A vertical soil pipe carrying excrement and liquid wastes.

stack

A general term used for any vertical line of soil, waste, or vent piping.

stack vent

A stack vent is the extension of a soil or waste stack above the highest horizontal drain connected to the stack.

trap

A fitting which prevents the passage of air, gas and vermin through a pipe without adversely affecting the flow of sewage or waste water.

unit vent

An arrangement of venting, installed so that one vent pipe will serve two traps.

vent

A pipe or opening used for circulating air in a plumbing system and for reducing the pressure on trap seals.

waste pipe

A pipe used to carry liquid wastes not containing excrement.

waste stack

A vertical pipe used to carry liquid wastes not containing excrement.

wet vent

That portion of a vent pipe through which liquid wastes flow.

Index

Y